Medical Ethics in Clinical Practice

Matjaž Zwitter

Medical Ethics
in Clinical Practice

 Springer

Matjaž Zwitter
Faculty of Medicine
University of Maribor
Maribor
Slovenia

Institute of Oncology
Ljubljana
Slovenia

Originally published in Slovenian language in 2018 by Cankarjeva založba
ISBN 978-3-030-00718-8   ISBN 978-3-030-00719-5   (eBook)
https://doi.org/10.1007/978-3-030-00719-5

Library of Congress Control Number: 2018964943

This Springer imprint is published by the registered company Springer Nature Switzerland AG
The registered company address is: Gewerbestrasse 11, 6330 Cham, Switzerland

Preface

In a concise phrase, the message of this book is:

how to be a good physician

Many have written about ethics: philosophers, theologians, journalists, politicians, economists, and others. This book about ethics is written by a physician. Day by day, year by year, physicians experience the distress of people who find themselves at the brink of life and death. The experience from real life is what separates mountaineers from television viewers and physicians from philosophers. This is not to say that the opinion of a physician counts more than that of a philosopher: the two simply observe a problem from different perspectives. When compared to other, more theoretical volumes on medical ethics, this book should contribute to a balanced understanding of the role that physicians play, or should play, in medicine and in society.

Medical Ethics in Clinical Practice is based on lectures to undergraduate students of medicine at the Faculty of Medicine in Maribor, Slovenia. Nevertheless, this is not a typical textbook: it takes us beyond a systematic analysis toward lived situations with no clear-cut distinctions or black and white categories. In ethics, values play a decisive role, and there are often no right or wrong opinions since everyone can assign different weights to different values. Thus, my goal was not to build a coherent and well-polished flawless system, but rather to spur and encourage creative thinking.

After the introductory chapter on the role of ethics in human relations, the text continues with a brief and admittedly simplified presentation of ethical theories and of ethical analysis, I attempted to avoid language that could only be understood by experts: my words are meant to stimulate practical reflection in everyone. The next few chapters deal with topics of importance for all physicians, regardless of their speciality: communication, interpersonal relations in a medical team, professional malpractice, and medicine with limited resources. In "the Portrait of a Physician," a broad landscape of our profession is presented, portraying exceptional historical figures as well as anonymous physicians, often overburdened and suffering from unrealistic expectations of their patients and of the whole of society. The last ten chapters present a wide array of ethical dilemmas in different branches of medicine, from preventive medicine, the beginning of life, and genetics to intensive care, end-of-life issues, and clinical research. By wandering through the paths outside of oncology as my own field of medical speciality, I here and there find myself walking on thin ice; however, this is a small price to pay, compared to the alternative of offering only general rhetoric without an opinion on concrete ethical dilemmas.

The final part of the book contains a list of ethical problems and tasks, as presented to students of medicine. I hope that the readers will take them as an encouragement for reflection. Some tasks—such as a visit to a historic monument, a Partisan Hospital from World War II—are not practical for readers outside of Slovenia. I kept them on the list to give an idea to other teachers of medical ethics about the wide spectrum of experience which can stimulate ethical discussion.

Clearly, a book of modest size cannot include all the ethical dilemmas of modern medicine. While I am most grateful to reviewers for their comments and suggestions, the responsibility for including or omitting particular topics and for interpretation of ethical dilemmas is mine. To keep the discussion short, I explained my attitudes toward each ethical issue and did not attempt to offer a deep and balanced discussion of each problematic medical activity. Together with a comprehensive list of references, such an extensive discussion would grossly increase the volume—with an inevitable decrease in the interest in reading the book.

While the book has been written with physicians and students of medicine as the target audience, I am confident that it will be of interest also to professionals and students of other professions devoted to human health, to journalists, and to a general audience with an interest in medicine.

Comments and opinions are welcome: matjaz.zwitter@guest.arnes.si

Sincere thanks:
- *To my wife, Neta, for many years of encouragement and support.*
- *To the dean, teachers, and students of the Faculty of Medicine in Maribor for the stimulating work environment.*
- *To Matjaž Lunaček, Marjan Kordaš, and Raanan Gillon for their critical comments and suggestions.*
- *To friends and colleagues who contributed illustrations or consented to their publication.*
- *To Kristijan Armeni and Terry Jackson for translation and language corrections and to Corinna Parravicini and Aruna R. Sharma and Vinodhini Subramaniam for valuable editorial support.*

Maribor, Slovenia   Matjaž Zwitter

Contents

Ethics and Law

1

Abstract

Human conduct and relations are regulated by legal and ethical norms. Why are there two systems? Legislation consists of clear formulations, is limited by political frontiers, and has clear sanctions. In contrast, ethical norms are broader and often limited to recommendations, may cross political or cultural borders, and may be applied retroactively. While the legal system adequately regulates relations among individuals of a comparable position, it often fails to protect the weak side in cases of significant differences in power and social weight. The relations of children to parents, students to teachers, citizens to politicians, readers to journalists, and patients to physicians are examples for which the legal system alone cannot offer adequate protection to the weak side, and for which ethical limitations to the power of the stronger side are essential.

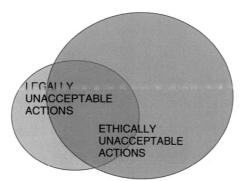

Partial overlap of legally unacceptable actions and ethically unacceptable actions. Some actions are not against the law but are ethically unacceptable; and vice versa

© Springer Nature Switzerland AG 2019
M. Zwitter, *Medical Ethics in Clinical Practice*,
https://doi.org/10.1007/978-3-030-00719-5_1

On his desert island, Robinson Crusoe needed neither laws nor ethics. However, when we live in a community with other people, our freedom is limited by the freedom, interests, and rights of others—the people we share our planet with and the people of the past and future generations, and also by the freedom, rights, and interests of other living beings. Therefore, we need rules that we can use as guidelines for avoiding conflicts. Such rules can be written as part of legal norms. As we are about to explore, however, legal norms alone are not enough: for a coherent and active society, we also need ethics.

1.1   Law

Law is based on explicit, written norms with a clear system for enacting such norms. Legal norms (approved international agreements, national constitutions, laws, and regulations) are organized hierarchically, should be mutually consistent, have a clearly defined spatial scope of their enactment, and can only apply to actions committed after the legal norm has been enacted. When an individual violates a specific legal norm, clearly defined procedures for establishing guilt and prescribed sanctions are in place.

Constitutions, laws, regulations, and other similar documents provide us with precise guidelines on how to act as individuals in a society. Why then do we need ethics?

1.2   Ethics

The foundations of ethics are different. When we evaluate an act as unethical, we most often do not refer to any concrete ethical document. For millennia, ethical rules have been incorporated into all religions. Fascinating are the many similarities between the Jewish and Christian ten commandments, the Islamic ten commandments, and the five principles of Buddhist ethics (prohibition of killing, stealing, lying, sexual misconduct, and alcohol consumption). It appears that the basic ethical principles are grounded in the fabric of human society and, therefore, extend beyond the limits of time and place set forth by political, national, or religious belonging.

Ethical commandments found in different religions apply to all members of a religious community. In a state where a religious doctrine is formally linked with state authorities or when it adopts a leading role, such commandments often apply to all citizens. In addition, smaller communities create their own specific ethical rules that are adopted by very different groups: physicians, nurses, teachers, scientists, journalists, economists, lawyers, retailers, artisans, politicians, athletes, hunters, and soldiers, to name just a few. All human pursuits have led to the creation of rules that should be respected if we want to perform the activities in line with the interests of other members of the group and the society as a whole. Ethical rules of any given group, therefore, act inwards and outwards: they regulate the relations within the group and protect the group as a whole from conflicts with a wider community.

Once we realize how variable the societal circles that adopt such ethical rules are, it becomes obvious that we are not talking about a coherent system, but rather a set of rules which often even contradict each other. Another difference between legal and ethical rules is that the latter do not generally prescribe sanctions; when ethical rules do include sanctions, these only carry "social" consequences—for example, an individual might be expelled from the group.

1.3   Law and Ethics: Why Do We Need Two Systems?

Why then do we need ethics in addition to legal norms for regulating our society? Would it not suffice if we simply translated some of the ethical provisions into the legal system?

Here one must realize that the law is fairly successful in regulating relations among individuals of comparable power. When the rules for regulating the relations are abused, the victim of comparable power has all legal means available for claiming his or her rights. The victim's situation changes drastically, however, when the involved parties happen to be in a relationship of clear inequality. Consider the following relationships: doctor–patient, politician–citizen, teacher–student, journalist–reader, parent–child, nursing staff–client in a retirement home. These are all instances of relationships with profound imbalances in knowledge, power, position in the local environment and in society at large, and in capacities for action, such that a potential victim cannot be sufficiently protected by the law. Moreover, the weaker party almost never relies on legal protection. Can you imagine citizens suing politicians for fraud when the latter fails to fulfill the promises that brought them to their current positions? In private conversations, patients will often complain about the way the healthcare staff treated them; nevertheless, filed complaints of this sort are relatively rare. Similar conclusions apply when considering the environmental damage we are leaving behind for our future generations: the children of our children cannot take legal action against us; nevertheless, they should have the right to be born in a healthy environment.

We have already seen that in cases of violations, ethics does not provide clear sanctions, as the law does. Let us point out another difference between the two systems: ethical guidelines are not as fixed as legal norms, they are more general and, as such, much more open to interpretation. Whereas laws apply only to activities after enactment of a legal norm and great attention is paid to how norms are put into words, ethical guidelines can be interpreted from a broad range of perspectives and can occasionally even be retroactive. Is this a disadvantage? I do not think so. In times of rapid changes, legal systems often cannot cope with the new aspects of all human activities. Consequently, numerous acts that we do not approve of and that are not beneficial to society are left unsanctioned. In business, financial markets, environmental damage, healthcare systems, especially when healthcare meets the pharmaceutical industry, there is a great deal of wrongdoing and fraud that the legal system has not (yet) recognized and included on the list of prohibited activities. Ethics aids us in assessing such acts and reaching a stance on the issues at hand.

Those who violate legal norms are subjected to the judicial system with prosecutors and judges. Violations of ethical principles, in contrast, are for the most part dealt within professional organizations, for example, in ethical councils operating within a medical chamber, bar association, or an association of journalists. It is true that there will be more room for compassion in professional circles when considering the misconduct of a colleague—the oft-voiced public criticisms of professional solidarity are therefore at least in part valid. However, those professional organizations that are aware of their societal role and responsibility will certainly not protect a single member to the detriment of all others and will hence clearly pronounce on violations of ethical rules. The goal should not be vengeance and punishment, but rather correcting mistakes. From the wrongdoer's perspective, an ethical conviction by professional colleagues—even if only in a moral sense—will have much more significant bearing than a conviction obtained by those who "do not know much about the topic anyway." Such concerns only strengthen the importance of ethical considerations.

In the chapters that follow, I focus on healthcare ethics, a part of bioethics as a broad field concerning all ethical issues relevant to biological sciences.

Suggested Reading

Arras JD, Fenton E, Kukla R, editors. The Routledge companion to bioethics. Abingdon: Routledge; 2015. ISBN-13: 978-0415896665.

Beauchamp TL, Childress JF. Principles of biomedical ethics. Oxford: Oxford University Press; 1979 (first edition)–2013 (seventh edition). ISBN: 9780199924585.

Gillon R. Philosophical medical ethics. Hoboken: Wiley; 1986. ISBN-13: 978-0471912224.

Grady C, Ulrich CM. Moral distress in the health professions. Berlin: Springer; 2018. ISBN-13: 978-3319646251.

Hope T, Savulescu J, Hendrick J. Medical ethics and law: the core curriculum. London: Churchill Livingstone; 2008. ISBN-13: 978-044310337.

Keown D. Buddhism: a very short introduction. Oxford: Oxford University Press; 2005.

Kerridge I, Lowe M, Stewart C. Ethics and law for the health professions. New South Wales: Federation Press; 2013. ISBN: 9781862879096.

Kuhse H, Singer P, editors. A companion to bioethics. Hoboken: Wiley-Blackwell; 2012. ISBN-13: 978-0631230199.

Moreno JD, Berger S, editors. Progress in bioethics. Cambridge, MA: MIT Press; 2012. ISBN: 9780262134880.

Ozolins JT, Grainger J, editors. Foundations of healthcare ethics: theory to practice. Cambridge: Cambridge University Press; 2015. ISBN-13: 978-1107639645.

Phalen RF. Core ethics for health professionals: principles, issues, and compliance. Berlin: Springer; 2017. ISBN: 978-3-319-56088-5.

Richards B, Louise J. Medical law and ethics: a problem based approach. Oxford, UK: Butterworths; 2013. ISBN: 9780409333534.

Rogers WA, Braunack-Mayer A. Practical ethics for general practice. Oxford: Oxford University Press; 2009. ISBN-13: 978-0-1992-3552-0.

Romanucci-Ross L, Tancredi LR. When law and medicine meet: a cultural view. Berlin: Springer; 2004. ISBN-13: 978-1-4020-6763-1.

Seay G, Nuccetelli S. Engaging bioethics: an introduction with case studies. Abingdon: Routledge; 2017. ISBN-13: 978-0415837941.

Steinbock B, editor. The Oxford handbook of bioethics. Oxford: Oxford University Press; 2007. ISBN-13: 978-0199562411.

ten Have H. Encyclopedia of global bioethics. Berlin: Springer; 2016. ISBN-13: 978-3319094823.

Veatch RM. The basics of bioethics. Abingdon: Routledge; 2016. ISBN-13: 978-1138425019.

Zaidi SH. Ethics in medicine. Berlin: Springer; 2014. ISBN-13: 978-3319010434.

Ethical Codes and Declarations

2

Abstract

From Hippocrates and Chinese and Islamic healers to modern times, ethical codes and declarations are an essential component of medical education and conduct. Even after many centuries, the early documents on medical ethics are outstanding examples of clear, precise, and timeless formulations. In recent decades, dramatic changes to the practice of medicine have led to an enormous increase in the volume of ethical declarations and documents. Nevertheless, these documents lag behind the rapid new developments in medical research and practice and cannot offer advice in situations in which some ethical costs are inevitable and in which a compromise has to be made.

© Springer Nature Switzerland AG 2019
M. Zwitter, *Medical Ethics in Clinical Practice*,
https://doi.org/10.1007/978-3-030-00719-5_2

Ο ΟΡΚΟΣ ΤΟΥ ΙΠΠΟΚΡΑΤΥ

Ομνυμι Απόλλωνα ίητρόν και Άσκληπιόν καιΥγείαν και Πανάκειαν και
θεούς παντός τε και πάσας, 'ίστορας ποιεύμενος, έπιτελέα ποιήσειν κατά
δυναμιν και κρίσιν έμήν βρκον τόνδε και ξυγγραφήν τήνδε· ήγήσασθαι μεν
τον διδάξαντα με την τέχνην ταύτην Τσα γενέτησιν έμοΤσι, και βίου
κοινώσασθαι, και χρεών χρηϊζοντι μετάδοσιν ποιήσασθαι, και γένος το εξ
ωϋτέου άδελφοϊς ίσον έπικρινέειν αρρεσι, και διδάξειν την τέχνην ταύτην,
ην χρηϊζωσι μανθάνειν, άνευ μισθοϋ και ξυγγραφής, παραγγελίης τε και
άκροήσιος και της λοιπής άπάσης μαθήσιος μετάδοσιν ποιήσασθαι υιοΤσί
τε εμοΓσι, και τοϊσι του έμέ διδάξαντος, και μαθηταϊσι συγ-γεγραμμένοισί
τε και ώρκισμένοις νόμω ίητρικω, αλλω δε ούδενί.

Διαιτήμασί τε χρήσομαι έπ' ώφελείτ) καμνόντων κατά δυναμιν και κρίσιν
έμήν, επί δηλήσει δε καϊ άδικίη ε'ίρξειν. Ου δώσω δε ουδέ φάρμακον
οϋδενι αιτηθείς θανάσιμον, ουδέ ύφηγήσομαι ξυμβουλίην τοιήνδε' ομοίως
δε ουδέ γυναικί πεσσόν φθόριον δώσω. Άγνώς δε και όσίως διατηρήσω
βίον τον έμόν και τέχνην την έμήν. Ου τεμέω δε ουδέ μην λιθιώντας,
εκχωρήσω δε έργάτησιν άνδράσι πρήξιος τήσδε.

'Ες οικίας δε όκόσας αν έσίω, έσελεύσομαι έπ' ώφελείτ) καμνόντων, έκτος
έών πάσης άδικίης έκουσίης καϊ φθορίης, της τε άλλης και αφροδισίων
έργων επί τε γυναικείων σωμάτων και άνδρώων, ελευθέρων τε και
δούλων. "Α δ' αν εν θεραπείη ή ίδω, ή ακούσω, ή και άνευ θεραπήίης κατά
βίον ανθρώπων, 'ά μη χρή ποτέ έκ-λαλέεσθαι ε'ξω, σιγήσομαι, άρρητα
ήγεύμενος εΤναι τα τοιαύτα.

"Ορκον μεν οδν μοι τόνδε έπιτελέα ποιέοντι, και μη ξυγχέοντι, είη
έπαύρασ-θαι και βίου καί τέχνης δοξαζομένω παρά πασιν άνθρώποις ες
τον αίεί χρόνον πα-ραβαίνοντι δε και έπιορκοϋντι, τάναντία τουτέων.

In the European tradition, the Hippocratic oath is the oldest and most respected code of medical
ethics

In the previous chapter, we mentioned several professional groups that recognize
the necessity of conforming to ethical activity, especially in situations of imbalance
of power. Among all professions, medical doctors were the first to recognize the
importance of a proper ethical approach.

There are two ways of determining whether an act is ethical. The first approach,
as presented in this chapter, is to compare the act against written guidelines—ethical
codes and declarations. In the chapters that follow, we will introduce ethical analysis as the second approach: the act is evaluated by reasoning about the degree to
which it is in accordance with specific ethical principles.

2.1   Hippocratic Oath and Other Ancient Documents

When discussing ethical codes and declarations, one has to begin with Hippocrates.
The most important legacy of Hippocrates and of his students is the Hippocratic
Oath. After 2500 years, this document remains the foundation of medical ethics.

Roughly 1000 years later, the renowned Chinese physician, Taoist, and alchemist Sun Simiao presented his ethical recommendations for physicians in his book entitled *On the Absolute Sincerity of Great Physicians* [1].

In the ninth century, Ishaq bin Ali Rahawi was the first to write a book on the principles of medical ethics in the Islamic world. In 20 chapters, the book lays out guidelines for the work of physicians. As the guardian of the patient's body and soul, the physician should work for the benefit of the patient, even if the patient is considered an enemy. The physician shall not prescribe any lethal drugs or assist with an abortion. Ali Rahawi also wrote the rules of good behavior, confidentiality and non-corruption, and mutual respect and solidarity between physicians [2].

Let us not forget that these documents written by physicians are older than any law aiming at regulating medical work and protecting patients' rights. At that point in time, ethics was already a major step ahead of law, and I dare to say the same still holds today.

Hippocratic Oath

I swear by Apollo Physician and Asclepius and Hygieia and Panaceia and all the gods and goddesses, making them my witnesses, that I will fulfill according to my ability and judgment this oath and this covenant:

- To hold him who has taught me this art as equal to my parents and to live my life in partnership with him, and if he is in need of money to give him a share of mine, and to regard his offspring as equal to my brothers in male lineage and to teach them this art—if they desire to learn it—without fee and covenant; to give a share of precepts and oral instruction and all the other learning to my sons and to the sons of him who has instructed me and to pupils who have signed the covenant and have taken an oath according to the medical law, but no one else.
- I will apply dietetic measures for the benefit of the sick according to my ability and judgment; I will keep them from harm and injustice.
- I will neither give a deadly drug to anybody who asked for it, nor will I make a suggestion to this effect. Similarly I will not give to a woman an abortive remedy. In purity and holiness I will guard my life and my art.
- I will not use the knife, not even on sufferers from stone, but will withdraw in favor of such men as are engaged in this work.
- Whatever houses I may visit, I will come for the benefit of the sick, remaining free of all intentional injustice, of all mischief and in particular of sexual relations with both female and male persons, be they free or slaves.
- What I may see or hear in the course of the treatment or even outside of the treatment in regard to the life of men, which on no account one must spread abroad, I will keep to myself, holding such things shameful to be spoken about.
- If I fulfill this oath and do not violate it, may it be granted to me to enjoy life and art, being honored with fame among all men for all time to come; if I transgress it and swear falsely, may the opposite of all this be my lot.

The Purpose of Medical Practice
1. The object is to help, not to gain material goods.
2. Save life and do not kill any living creature.
3. Do not seek fame: virtuous conduct will be rewarded by humans and spirits.

The Requirements of a Great Physician
4. Master the foundations of medicine thoroughly, work energetically and unceasingly.
5. Be mentally calm and firm in disposition; do not give way to selfish wishes and desire.
6. Commit oneself with great compassion to save every living creature.

Manner of Medical Practice
7. Possess a clear mind and maintain a dignified appearance.
8. Do not be talkative, engage in provocative speech, or make fun of others.
9. Do not ponder upon self-interest and fortune; sympathize and help wholeheartedly.
10. Examine and diagnose carefully, prescribe accurately, and cure effectively.

Attitude Towards Patients
11. Treat everyone on an equal basis, no matter whether they are rich or poor.
12. Do not reject or despise a patient who suffers from abominable diseases such as ulcers and diarrhea: be compassionate and sympathetic.
13. Do not enjoy oneself in a patient's house while the patient is suffering.

Attitude Towards Other Physicians
14. Do not belittle another physician in order to exalt one's own virtue.
15. Do not discuss others and decide about their rights and wrongs.

Why are we mentioning the Hippocratic Oath, Sun Simiao's ethical guidelines, and Rahawi's recommendations? Despite the fact that these texts were created in different cultural, religious, and social environments over a span of more than 1500 years, there are astonishing similarities between them. Diligence and professional competence, protection of life, compassion and help for the sick regardless of their social status, humility and relinquishing one's own comfort, confidentiality, solidarity among physicians—it is remarkable how valid these written rules remain today.

2.2   Medical Ethics in Modern Documents

Nowadays, written ethical codes and declarations abound. Such documents have been adopted by national and international professional medical associations, as well as by specialized medical associations in the fields of gynecology, psychiatry, pediatrics, emergency medicine, intensive care medicine, public health, occupational medicine, and sports medicine, to name just a few. Due to their high number, physicians cannot be familiar with all the documents and possibly none of us has a complete overview. The second negative consequence of the abundance of ethical codes and declarations is contradictory information across documents. Even more so: one can easily spot inconsistencies and internal contradictions within a single document. The third drawback is that written recommendations are constantly lagging behind the developments in medicine. Think about research in genetics, the ever more intensive involvement of preventive medicine into the healthy population, development in medical diagnostics and treatment, and the high degree of entanglement between contemporary medicine and the interests of the pharmaceutical industry—these are all areas where both legislation and ethical codes and declarations fall behind recent developments.

2.3 Advantages and Disadvantages of Normative Ethics

It would be of course wrong to suggest that we could do away with medical laws, codes of medical ethics, or the declarations of the World Medical Association. In its 2500 years, the oldest of the documents—the *Hippocratic Oath*—has not lost its significance. In many respects, codes and declarations provide good general guidelines. However, one would be misled into thinking that the sheer existence of an ethical code would help us solve numerous ethical dilemmas in physician's work. Real ethical dilemmas arise when each of the measures physicians have at their disposal incurs some ethical costs. To turn the phrase differently: genuine dilemmas are those for which one has to choose among several actions, of which none is without unwanted consequences. Ethical codes and declarations provide us with no advice for such situations. This leads us to the next chapter: ethical theories and ethical analysis.

References

1. Fu-Chang Tsai D. Ancient Chinese medical ethics and the four principles of biomedical ethics. J Med Ethics. 1999;25:315–21. http://www.ncbi.nlm.nih.gov/pmc/articles/PMC479240/pdf/jmedeth00005-0025.pdf
2. http://muslimheritage.com/topics/default.cfm?ArticleID=570

Suggested Reading

Askitopoulou H, Vgontzas AN. The relevance of the Hippocratic Oath to the ethical and moral values of contemporary medicine. Part I: The Hippocratic Oath from antiquity to modern times. Eur Spine J. 2018;27:1481–90. https://doi.org/10.1007/s00586-017-5348-4.
Askitopoulou H, Vgontzas AN. The relevance of the Hippocratic Oath to the ethical and moral values of contemporary medicine. Part II: Interpretation of the Hippocratic Oath-today's perspective. Eur Spine J. 2018;27:1491–500. https://doi.org/10.1007/s00586-018-5615-z.

Caporale C, Pavone IR. International biolaw and shared ethical principles: the universal declaration on bioethics and human rights. Abingdon: Taylor & Francis; 2017. isbn:9781472483980.
Malik AY, Foster C. The revised declaration of Helsinki: cosmetic or real change? J R Soc Med. 2016;109:184–9. https://doi.org/10.1177/0141076816643332.

Selected Codes of Medical Ethics

Argentina: https://www.ama-med.org.ar/images/uploads/files/ama-codigo-etica-castellano.pdf
Australia: https://ama.com.au/media/new-code-ethics-doctors
Bahrain: http://www.nhra.bh/files/files/2017/HCP_DOC/Standards/Standards%20latest/code%20 of%20medical%20ethics%20v1%204.pdf
Bangladesh: http://www.readcube.com/articles/10.3329/birdem.v4i1.18544
Brasil: http://www.cremers.org.br/pdf/codigodeetica/cem_e_cpep.pdf
Canada: https://www.cma.ca/Assets/assets-library/document/en/advocacy/policy-research/CMA_ Policy_Code_of_ethics_of_the_Canadian_Medical_Association_Update_2004_PD04-06-e. pdf
Egypt: https://srhr.org/abortion-policies/documents/countries/04-Egypt-Code-of-Medical-Ethics- Ministry-of-Health-and-Population-2003.pdf
Ethiopia: http://www.emacpd.org/sites/default/files/resource_center/Medical%20Ethics.pdf
Europe: Patuzzo S, Pulice E. Towards a European code of medical ethics: ethical and legal issues. J Med Ethics. 2017;43:41–6.
France: https://www.conseil-national.medecin.fr/sites/default/files/code_de_deontologie_version_ anglaise.pdf
Germany: https://www.bundesaerztekammer.de/fileadmin/user_upload/downloads/MBOen2012. pdf
Hong Kong: https://www.mchk.org.hk/english/code/files/Code_of_Professional_Conduct_2016. pdf
India: https://www.mciindia.org/CMS/rules-regulations/code-of-medical-ethics-regulations-2002
Italy: https://portale.fnomceo.it/wp-content/uploads/2018/03/CODICE-DEONTOLOGIA- MEDICA-2014.pdf
Japan: https://www.med.or.jp/english/about_JMA/principles.html
Nepal: http://www.nmc.org.np/information/nmc-code-of-ethics.html
New Zealand: https://www.nzma.org.nz/__data/assets/pdf_file/0016/31435/NZMA-Code-of- Ethics-2014-A4.pdf
Nigeria: http://www.mdcnigeria.org/Downloads/CODE%20OF%20CONDUCTS.pdf
Norway: http://legeforeningen.no/Om-Legeforeningen/Organisasjonen/Rad-og-utvalg/ Organisasjonspolitiske-utvalg/etikk/Code-of-Ethics-for-Doctors-/
Pakistan: http://www.pmdc.org.pk/LinkClick.aspx?fileticket=v5WmQYMvhz4%3D&tabid=292 &mid=845
Philippines: https://www.philippinemedicalassociation.org/wp-content/uploads/2017/10/FINAL- PMA-CODEOFETHICS2008.pdf
Poland: http://www.nil.org.pl/english/from-the-medical-code-of-ethics
Saudi Arabia: https://www.iau.edu.sa/sites/default/files/resources/5039864724.pdf
Singapore: http://www.healthprofessionals.gov.sg/content/hprof/smc/en/topnav/guidelines/ethi- cal_code_and_ethical_guidelines.html
Slovenia: https://www.zdravniskazbornica.si/docs/default-source/zbornicni-akti/code_of_medi- cal_ethics.pdf?sfvrsn=879c2836_2
Spain: https://web.archive.org/web/20120322030730/http://www.actasanitaria.com/fileset/ doc_65737_FICHERO_NOTICIA_59946.pdf

Trinidad and Tobago: http://www.mbtt.org/CodeOfEthics_Responsibilities_to_profession.htm
UK: https://www.gmc-uk.org/-/media/documents/good-medical-practice%2D%2D-english-1215_pdf-51527435.pdf
USA: https://www.ama-assn.org/delivering-care/ama-code-medical-ethics
World Medical Association: https://www.wma.net/policies-post/wma-international-code-of-medical-ethics/

Ethical Theories

3

Abstract

Our brief and admittedly incomplete discussion will be limited to four ethical theories: utilitarian ethics, deontological (or Kantian) ethics, virtue ethics, and principlism. As the oldest of the three, utilitarian ethics is based on the ethical principle of beneficence: with the available resources, do as much good as you can. This advice may be reasonable in a massive disaster or in preventive medicine, but the application of pure utilitarian ethics to regular physicians' activities with patients might lead to very problematic decisions, such as the legalization of euthanasia and denial of the right to decent palliative care. Deontological, or Kantian, ethics is based on respect of autonomy and sacredness of every human being. Once we agree on an ethical rule, it should be applied regardless of the consequences. Nevertheless, it is hard to argue that consequences, specific circumstances, or personal relations are irrelevant for a physician's decision. As the third ethical theory, virtue ethics focuses on virtues (or vices) of the moral agent and is associated with a good, happy, flourishing life (*eudaimonia*). Since excess wealth does not contribute to happiness, the most esteemed virtues are those that serve others. Blind virtues not supported by experience may be harmful, and practical wisdom (*phronesis*) has been proposed as a link between virtues and the consequences of every human activity. Finally, we will present principlism, as based on four ethical principles: autonomy, beneficence, non-maleficence, and justice and as a framework for the ethical analysis of complex situations.

© Springer Nature Switzerland AG 2019
M. Zwitter, *Medical Ethics in Clinical Practice*,
https://doi.org/10.1007/978-3-030-00719-5_3

ANTOINE DE SAINT-EXUPÉRY

Le Petit Prince

Due to his faithfulness, compassion and veracity, Le Petit Prince, a creation of Antoine de Saint-Exupéry, is a most convincing and charming embodiment of virtue ethics

Let us repeat the final words of the previous chapter: real ethical dilemmas are those for which none of the possible actions is without ethical costs. Here is a very simple example: if in my office I spend an hour talking to a severely distressed patient who urgently needs the conversation, I have to let other patients wait; however, they might need urgent treatment as well.

Ethical theories tend not to provide concrete solutions. Rather, they are a tool with which everyone can approach a dilemma. Ethical theories help us to evaluate possible actions and thus may lead to a recommendation.

Each ethical theory could easily provide enough material for an entire book of its own. Our discussion will remain brief and admittedly incomplete: a comprehensive theoretical discussion on ethical theories would be beyond the scope of this book; it would also require a different author. As a physician, I simply cannot pretend to be in a position to write long philosophical treatises. I will, therefore, leave aside many important ethical theories and provide sketches of only four theories. We will begin by discussing three that are necessary for further discussion: utilitarian ethics, deontological (Kantian) ethics, and virtue ethics. As we will soon see, these three ethical theories are not sufficient, which is why we will conclude the chapter by introducing the ethics based on four ethical principles, sometimes called common morality ethics or principlism.

3.1   Utilitarian Ethics

The basic principle of utilitarian ethics is *do good*, which is the oldest and most straightforward ethical principle. We can rephrase this principle in a way to make it useful for solving concrete ethical dilemmas as follows:

> Using the available means, act in such a way as to do as much good as possible.

This sounds simple and useful. When is it appropriate to apply this principle and when not?

Principles of utilitarian ethics are useful primarily in catastrophic situations. In times of war, earthquakes, or other major accidents, we have to save as many lives as possible in the shortest amount of time with very limited means. In a rational triage, one would group the wounded into those with minor wounds and injuries who can wait and those who require that we urgently stop their bleeding, clear their airways or prevent further aggravation of their health condition.

Utilitarian ethics also offers appropriate advice to activities in the field of public health and preventive medicine. The goals of these branches of medicine are often of statistical nature, rather than directed towards an identifiable individual. When choosing among different activities, we will often choose those with the greatest benefit for the available resources.

However, a large part of medicine and medical work does not consist of catastrophic situations or public health. Despite the fact that we encounter the problem of limited means on a daily basis, it would be very inappropriate and dangerous to rely on utilitarian ethics in everyday medical work. Let us take an example of the utilitarian pleading of euthanasia found on the web:

> In the financial aspect, the legalization of euthanasia will reduce the burdens for both the patients' families and the medical system. Because of taking care of the patients with long-term disease, countless families have spent a lot of time and paid most of their fortune. The medical system also has to provide a good quantity of resources. Besides, in many countries, there is a shortage of hospital space and the energy of doctors. If euthanasia can be conducted, the resources could be used for people whose lives could be saved instead of those who are not given hope to recover [1].

Indeed, in utilitarian reasoning, the argument in favor of euthanasia holds: in terms of public health and healthcare statistics, care for the terminally ill provides much less benefit than some other branches of medicine. Nevertheless, many—probably most—physicians believe that one cannot simply adopt euthanasia of the terminally ill to reallocate the resources for vaccination or, for example, to purchase a helicopter for urgent medical transport.

Let us for a moment digress and consider a situation outside medicine in which utilitarian ethics is similarly of limited use. Imagine the police are hunting a terrorist group leader responsible for the deaths of hundreds of people, a man who is planning new attacks on civilians. The police decide to capture the family of the terrorist. They broadcast a video of the terrorist's 12-year-old son and announce they will cut off one finger from the boy's hand every day—as long as the terrorist does not surrender himself. Even a most hardened criminal would probably not be able to stand by and watch his child being mutilated and would surrender. Adopting a utilitarian perspective, one could conclude: for the price of a child's two fingers, we captured a terrorist who could kill many more people in the coming days. Even if true, it should be clear that we should not involve the terrorist's family in the hunt.

It would be wrong to entirely abandon utilitarian ethics. Apart from catastrophic situations and public health, rational use of limited means is also needed in everyday life and should be taken into account in our decision-making. It is essential to emphasize, however, that an act should not be valued only by considering its consequences, and that ethics cannot be based solely on the principle of *doing good*, let alone on the principle of maximizing good.

3.2   Deontological Ethics

Whereas from the perspective of utilitarian ethics, the ethical status of an act depends only on the consequences, deontological or more specifically Kantian ethics is based on the opposite principle: an act can be deemed ethical only insofar as it complies with an ethical accord or an a priori agreed-upon ethical norm. Therefore, if we agree and accept that "Do not lie!" is an ethical norm, then lying can never be ethical even if by lying one could prevent severe negative consequences. According to Kant, then, it is not the consequences of the act that can be deemed good and bad; what is important is that the act stems from genuine ethical motivation. Kant concluded that everyone should always act in line with what he or she wants to be a general, permanently valid rule. Finally, a human being, as a unique being, can never be only a means; he or she can only be the goal of our actions.

In utilitarian ethics, the most prominent status is given to the principle of beneficence. Kantian ethics, however, leads us to appreciate the individual, one's free will, and rational responsibility for one's decisions—the most important is, therefore, the ethical principle of the respect for autonomy. By requiring one to follow, without exception, universal rules, Kant excluded emotions and interpersonal relations from entering ethical reflections. Even more: an act is ethical only if the motivation comes from the general rule. If done from fear or love, the ethical value of the same act is neutral. In Kantian ethics, there is no place for compromise and no weighing of motivations for acting against the consequences, regardless of how predictable, and thus avoidable, the latter could be.

With all due respect to the great philosopher, I nevertheless wonder how someone can claim that attachment, love, and responsibility for a fellow human being must not influence our decisions? The man who never had a family, who lived in his own world of philosophical discussions, who never left his hometown of Königsberg (now Kaliningrad) on the cold Baltic Sea gave us a rigid and cold system of advice for life. Please forgive me if I say that I can find more useful and real-life philosophy in *The Little Prince* by Antoine de Saint-Exupéry. Do you recall the fox and the explanation of why the Little Prince's rose is more important and dearer to him than a whole bush full of seemingly equal roses?

No, in developing ethical guidelines, we cannot single out our emotions, interpersonal relationship, and the consequences of our actions. The autonomy of the individual is a very important ethical principle; however, it cannot serve as our sole criterion for judging whether an act can be deemed ethical or not. This brings us first to virtue ethics, and then to the four ethical principles.

3.3 Virtue Ethics

In ethical assessment, utilitarians look at the consequences of a particular act, and deontological ethicists at the conformity of the act with pre-defined rules. Virtue ethics stands in a different position: its foundation is the ethical characteristics of a person. Certain personal characteristics are recognized as virtues and others as vices. Honesty, generosity, compassion, courage, justice, fidelity, and veracity are virtues that characterize an ethical person, while deception, selfishness, cruelty, infidelity, and disingenuousness are some of the many characteristics that denote a vicious person.

Before attributing certain virtues or vices to a person, some limitations have to be considered. No opinion can be given about persons whom we do not know well, or whose motives are not clear. Beneficence towards another person may be sincere and altruistic or may be based on fear or selfishness. In the first case, we are observing an act of a virtuous person; in the second case, the act may be ethically neutral or even wrong. We also have to consider that virtues are often less than ideal: many people are beneficent, honest, and compassionate to a certain level but are not perfect. In addition, someone may be a highly virtuous person in many respects but have a blind spot, such as an ethically unacceptable harsh position towards immigrants.

While virtue ethics is not based on consequences of actions, a virtuous person will foresee these when considering a particular action. An essential part of virtue ethics is, therefore, a rational ability to discern, also called practical wisdom or, in Greek, *phronesis*. A child or an adolescent often acts with the best of intentions yet it is not able to foresee possible bad consequences of his act. He lacks practical wisdom which only comes after years of life experience. It is only an experienced person who decides to help an old beggar but does not offer money to a young drug addict, one who would immediately spend it on a new shot of heroin. Only an experienced physician will be able to decide whether to reveal the full truth to a severely ill patient and accept the risk of the patient's severe depression, or whether conceal-ing the truth would be an unacceptable deception.

Virtue ethics is associated with the concept of a good, happy, flourishing life—in Greek *eudaimonia*. According to this concept, a happy life is not one devoted to selfish striving for wealth since an excess of wealth does not contribute to happi-ness. On the contrary, excessive accumulation of material goods brings insecurity and will often dilute one's inner feeling of happiness. For true, deep happiness, it is essential to re-direct one's attention and develop virtues that will serve others—one's relatives and friends, a wider circle of people, or (for those with special abili-ties) the whole of humanity. It is clear that the degree of happiness also depends on some other external factors and on fortunate or unfortunate circumstances which are beyond our control. The elementary postulate holds: only those who unselfishly develop and nourish their virtues may attain deep inner happiness; and on the oppo-site side, vicious persons will never experience real happiness and will never flour-ish, no matter how rich and mighty they may be.

Virtue ethics may be seen as an upgrading of utilitarian and deontological ethics. While virtues are in a central position, the incorporation of practical wisdom also brings consequences into virtue ethics. The rules of deontological ethics are impor-tant, yet not unbreakable. Finally, the main advantage of virtue ethics over the other two theories is that it includes human relations, and indeed gives a central position to them, a factor that has a decisive role in the practical reasoning of all of us.

A long-lasting relationship of a physician with a chronic patient, relationships within a family, a community of friends, or a healthcare team depend less on rules and more on trust and fidelity. In the absence of such trusting relationships, such as the first visit of a new patient, rules and codes of conduct come to the forefront. Thus, virtue ethics may be of lesser importance in contacts with strangers or in situ-ations when the decisions are less personal. In such instances, the four ethical prin-ciples and ethical analysis might be the preferable approach.

3.4   The Four Principles and Common Morality Ethics

Four decades ago, Tom Beauchamp and James Childress presented their ethical theory, which is not based on a single principle. Later also called *principlism*, their theory rests on four basic principles:

- Autonomy of the individual
- Non-maleficence
- Beneficence
- Justice

Autonomy of the individual is more than a simple right to self-determination. One can state that this principle is the first among equals and has, at least in our European and North American culture, such a prominent status that we shall dedicate an entire chapter to it. At this point, let us briefly point out that the autonomy of the individual includes the right to being informed (for there is no autonomy without information) and the right to confidentiality and privacy (otherwise other parties might influence an individual's decisions without considering his or her will). The respect for individuals' autonomy depends to a greater extent than other principles on the cultural environment: in southern and eastern parts of the world, the values and interests of the community are treasured significantly more than those of the individual. In the next chapter, we will also discuss the questions pertaining to limited autonomy: surrogate decisions and advance directives.

When reviewing the basics of utilitarian ethics, we learned about the ethical principle of *beneficence*; the principle of *non-maleficence* is closely related. Some ethicists prefer to subsume both of these principles under a single principle of beneficence. However, for our purposes, and especially for the purposes of ethical analysis, it is nevertheless important to treat the two principles separately. The difference between beneficence and non-maleficence is in the spectrum of people affected by my actions, be it good or bad. The respect for beneficence applies to a much narrower circle of people than the principle of non-maleficence does. The commitment to beneficence holds much more strongly for my relationships with my family members, friends, high-school classmates, my patients, rather than for perfect strangers. Even when I am off duty on a Saturday, I will go to the hospital and see the patient to whom I prescribed a risky treatment the day before; but I will let my colleague on duty take care of other patients who are not "mine." Similarly, if my brother loses his house due to a fire, I will welcome him, his wife, three kids, and their dog in my house for the time they need to find a new place. On the other hand, it would likely appear to my wife very strange and hard to understand if I wanted to host five homeless strangers in our apartment.

The principle of *non-maleficence* is distinguished from the principle of beneficence by the circle of people to whom it applies. Essentially, here the circle of people has no limits. We must respect the principle of non-maleficence even in our relationships with complete strangers. No one has the right to spit at a stranger's face, to spur hate or disseminate racist thoughts, or to destroy the environment for the future generations.

Justice is the fourth ethical principle. Here, we will be concerned only with the principle as it applies to medicine and will leave aside the issue of justice in everyday life and in relations outside of healthcare. When it comes to disease, justice can be understood in two very different ways. If we oversimplify the current global image, there is, on the one hand, the European conception of justice in medicine and, on the other, medical justice as perceived in the USA and many developing countries. In Europe, Canada, Australia, New Zealand, and some other countries, access to healthcare is based on solidarity and is not limited by the individual's social status and his or her past contributions to the healthcare fund. Protection of health is, therefore, a human right and is guaranteed constitutionally and by law. Health insurance normally includes all citizens. It is true, however, that treatment is often not completely free. Most of the healthcare systems require partial participation in costs so as to prevent misuse of the system. Nevertheless, such participation does not represent a huge financial burden for people, and even the very expensive modes of treatment are accessible to those from the lower social strata.

In countries with no universal healthcare, justice is conceived differently: anyone is free to decide what kind of health insurance he or she wants to purchase. It would be considered by many to be unfair or unjust to use the money collected from those who have paid insurance premiums also to provide medical care for those who have not paid for medical insurance. For those without health insurance, many countries will cover the costs of life-saving treatment or the basic healthcare for the poorest. Here, the solidarity ends. Roughly one quarter of middle-class Americans are without healthcare insurance, and even more are those for whom their insurance does not cover costs for very expensive modes of treatment.

This concludes our overview of the four principles of medical ethics. In the next two chapters, we will present the concept of moral status and then connect the four principles in the context of ethical analysis.

Reference

1. http://lang-8.com/48307/journals/156653

Suggested Reading

Gillon R. Defending the four principles approach as a good basis for good medical practice and therefore for good medical ethics. J Med Ethics. 2015;41:111–6. https://doi.org/10.1136/medethics-2014-102282.

Gómez-Lobo A, John Keown J. Bioethics and the human goods: an introduction to natural law bioethics. Washington, DC: Georgetown University Press; 2015. ISBN-13: 978-1626161634.

Heubel F, Biller-Andorno N. The contribution of Kantian moral theory to contemporary medical ethics: a critical analysis. Med Health Care Philos. 2005;8:5–18.

Hursthouse R, Pettigrove G. In: Zalta EN, editor. Virtue ethics, the Stanford encyclopedia of philosophy. Winter 2016 ed. https://plato.stanford.edu/archives/win2016/entries/ethics-virtue/

Mandal J, Ponnambath DK, Parija SC. Utilitarian and deontological ethics in medicine. Trop Parasitol. 2016;6:5–7. https://doi.org/10.4103/2229-5070.175024.

Stapleton G, Schröder-Bäck P, Laaser U, Meershoek A, Popa D. Global health ethics: an introduction to prominent theories and relevant topics. Glob Health Action. 2014;7:23569. https://doi.org/10.3402/gha.v7.23569.

Sumner LW, Boyle J, editors. Philosophical perspectives on bioethics. Toronto: University of Toronto Press; 1996. ISBN-13: 978-0802071392.

Svenaeus F. Phenomenological bioethics: medical technologies, human suffering, and the meaning of being alive. Abingdon: Routledge; 2018. ISBN-13: 978-1138629967.

van Zyl L. Virtue ethics: a contemporary introduction. Oxford and New York: Routledge; 2018. ISBN: 978-0415836166.

Moral Status

4

Abstract

Assignment of moral status to those concerned is an essential element of every ethical deliberation. Indeed, differences in definition of the moral status of a fertilized human egg cell, of an embryo, of a patient in the terminal phase of an incurable disease or of a laboratory animal are at the roots of ethical disputes. Several criteria for definition of moral status have been proposed, including human properties, cognitive properties, moral agency, sentience, and ability to establish relations. Since none of these criteria adequately covers all dimensions of moral status, only a combination of them will lead to a balanced definition. The question of moral status should not be understood as an all-or-nothing phenomenon, but rather as a gradual acquisition of moral status, depending both on the degree of reflection of the individual and on her/his future potential.

© Springer Nature Switzerland AG 2019
M. Zwitter, *Medical Ethics in Clinical Practice*,
https://doi.org/10.1007/978-3-030-00719-5_4

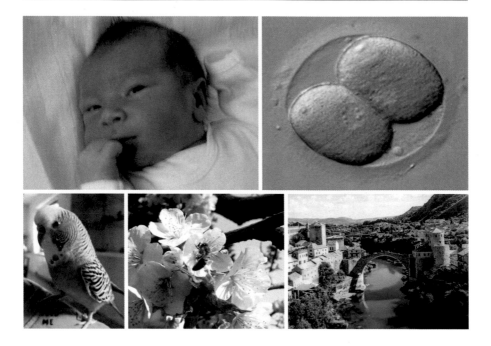

A newborn, a fertilized human egg cell, a pet, a bee, a monument. Moral status should not be limited to human species and is a matter of degree, rather than an all-or-nothing attribute

In discussions of ethics, we speak about *moral status*: only those who have moral status have the right to expect that others will respect their ethical rights. Indeed, the definition of moral status opens the doors to some of the most critical ethical discussions. Diverging views on the right to abortion, on procedures linked to physician-assisted insemination, on euthanasia for terminally ill patients or for individuals in persistent vegetative state, on vegan diets, or on our commitment to sustainable development with the preservation of the planet for future generations—all these discussions sooner or later touch upon the question of the moral status of the individuals concerned. Religions, traditional relations, as well as financial and economic aspects play a role in defining whom to protect. As an example, consider the recent debate on global ecology: do we assign moral status to the children of our grandchildren, those to be born sometime in the middle of the twenty-first century? Or may we act as if we had no responsibility for their lives or for their future problems?

4.1   Who Has Moral Status?

Let us start with a list of those to whom moral status might be assigned:

- Adult competent person
- Woman in a country with legislation based on Islam

- An illegal migrant on a boat to Italy
- 12-year-old boy working in a mine in Africa
- 3-year-old child
- Patient in persistent vegetative state
- Neonate
- Fetus in the 32nd week of pregnancy
- Fetus in the 8th week of pregnancy
- Fertilized egg cell
- A dead person
- A child who will be born in 2030
- A chimpanzee
- A laboratory mouse
- A cow
- A whale
- A mosquito
- "Yaksugi": a Japanese cedar tree more than 1000 years old and under UNESCO World Heritage protection, on the Yaku Island in Japan
- Palmyra, an ancient Semitic city in Syria

There should be no doubt that all of us with a capability of deciding about ourselves have moral status. Still, in some parts of the world, women continue to have lower moral status in comparison to men. Not long ago, moral status was not granted to slaves. While slavery has been prohibited, relations identical to slavery persist in many parts of the world, here including some affluent countries which deny basic human rights to immigrant workers. When politicians discuss the problem of illegal migrations, they seem to accept the fact that thousands of migrants continue to drown during the unsuccessful crossing of the Mediterranean. Imagine that 10,000 European citizens were trapped on the shores of North Africa: European governments would immediately send planes or ferries and bring them quickly and safely across the sea. Clearly, the lives of African or Asian migrants are not considered worth the same effort and, in practice, they are not considered to be persons with full moral status.

4.2   Criteria for Moral Status

For some philosophers, recognition of moral status is limited to persons capable of self-reflection. In a heavily criticized treatise, Peter Singer denied the moral status of newborns, thus justifying infanticide in cases in which the existence of the neonate was incongruent with the wishes of the parents [1]. It is clear that such views are unacceptable.

What is the moral status of an unconscious person, especially if in a terminal stage of an incurable disease? What is the moral status of a fetus, of a fertilized human egg cell? What is the moral status of a dead person? Has a chimpanzee moral status and, if so, is it higher than that of a frog or a mosquito?

A comprehensive presentation of the wide philosophical field on moral status is clearly beyond the purpose of this book. In their classic text, Beauchamp and Childress present five theories of moral status. According to a *theory based on human properties*, all humans have full moral status, and only humans have moral status. A *theory based on cognitive properties* does not limit the moral status to *homo sapiens*; rather, moral status belongs to individuals capable of perception, memory, understanding, and thinking. Next is the *theory based on moral agency*: moral status is attributed to those individuals with the capacity to act as moral agents, capable of making moral judgments, and with motives that can be judged morally. Even wider is the definition of moral status according to *a theory based on sentience*, according to which the key criterion is consciousness with the capacity to suffer and to feel pain or pleasure. Finally, *a theory based on relationships* emphasizes the importance of social interaction and reciprocity.

It is difficult to take a single one of the five theories as an instrument to draw a line between those individuals with moral status and those who are not granted moral status and, hence, do not have rights. Such a position would deny moral status to non-human animals such as great apes (using the theory based on human properties), or to fetuses and unconscious patients (using theories based on cognitive properties and on moral agency). Most of us, therefore, believe that for the definition of moral status, a blend of the theories above is the preferred approach. It is also clear that grading of moral status has to be introduced: while I accept that my neighbor's dog has some degree of moral status, it is still lower than the moral status of a child, and I may justly kill the dog if it attacks a child. The question of moral status should not be understood as an all-or-nothing phenomenon, but rather as a gradual acquisition of moral status, depending both on the degree of reflection of the individual and on his future potential. According to this view, a dog has a higher moral status than a fish in an aquarium, and a fetus in the 4th month of pregnancy has a higher moral status when compared to a fertilized egg cell.

The criteria of human properties, cognitive properties, moral agency, sentience, and relationship all coupled with grading—with these, have we reached a widely accepted approach to the question of moral status? We should understand that there will never be a consensus on the weight of individual criteria, let alone on grading. Many factors influence our reasoning, including religion, traditional beliefs, and the degree to which individual autonomy is limited by the interests of society. In addition, most discussions on moral status deal with the moral status of individuals, either of human or of non-human origin. Yet, moral status should also be attributed to groups of individuals with no identifiable individual names, such as refugees from Afghanistan, or members of the community of Jehovah's witnesses. Potential moral status should be given to future generations of humans and of all living beings, so they too should be included in the ethical analysis.

What about the dead? They have no cognitive properties, cannot be considered to be moral agents, and have no sentience; yet, they are human, and they are linked to us according to the theory of relationship. Regardless of religious or cultural viewpoints, we all respect our ancestors, remember them, and often feel that they continue to be with us, so they too deserve a degree of moral status.

We started this chapter with a list of those to whom moral status might be assigned. At the end of the list, I added a rare tree and a historic monument. In my view as a humble physician, destroying such a tree or a monument is not only illegal but is also highly immoral. Yet, using the five theories mentioned before, I see no possibility of assigning them moral status. I, therefore, conclude this short chapter with a plea to fellow philosophers: please invent a way to assign moral status also to miracles of nature and to monuments of human history.

Reference

1. Singer P. Discussing infanticide. J Med Ethics. 2013;39:260.

Suggested Reading

Berglund C. Ethics for health care. Australia: Oxford University Press; 2012. ISBN: 9780195519570.

Dilley S, Palpant NJ. Human dignity in bioethics: from worldviews to the public square. Abingdon: Routledge; 2013. ISBN-13: 978-1138922198.

Gross D, Tolba RH. Ethics in animal-based research. Eur Surg Res. 2015;55:43–57. https://doi.org/10.1159/000377721.

Jiwani B. Clinical ethics consultation: a practical guide, SpringerBriefs in Ethics. Berlin: Springer; 2017. ISBN: 978-3-319-60376-6.

Paulson S, Berlin HA, Miller CB, Shermer M. The moral animal: virtue, vice, and human nature. Ann N Y Acad Sci. 2016;1384:39–56. https://doi.org/10.1111/nyas.13067.

Pellegrino ED, Schulman A, Merrill TW, editors. Human dignity and bioethics. Indiana: Notre Dame Press; 2009. ISBN-13: 978-0268038922.

Serna P, Seoane J-A. Bioethical decision making and argumentation. Berlin: Springer; 2016. ISBN-13: 978-3319434179.

Ethical Analysis

5

> *Real life is, for most men, ... a perpetual compromise between the ideal and the possible*
>
> Bertrand Russell

Abstract

Real ethical dilemmas are those for which every possible action (including the omission of action) brings some ethical costs to a person, or to a group of persons involved. Ethical analysis begins with a list of persons involved and then explores, for each of them, the ethical benefits and costs of a certain action according to the four ethical principles: autonomy, beneficence, non-maleficence, and justice. In this way, two or more possible actions with their ethical benefits and costs are compared. Ethical analysis does not automatically lead to the recommended action; rather, it is a practical approach to analyze a dilemma, identify the problematic aspects of each of the possible actions, and alleviate the related ethical costs.

© Springer Nature Switzerland AG 2019
M. Zwitter, *Medical Ethics in Clinical Practice*,
https://doi.org/10.1007/978-3-030-00719-5_5

In ethical analysis, two or more actions with their ethical benefits and burdens are compared

If an act exists that will benefit all concerned and will not offend anybody, including oneself, then ethics or ethical analysis is not needed. Ethical consideration is also not applicable to situations in which the actor has no choice. Very often, however, we do have a choice; nevertheless, each act brings some benefits and some burdens to the persons whom the act involves. Ethical analysis is helpful in such situations.

The three steps of ethical analysis are:

- A list of possible actions
- List of persons or groups of persons whom the actions concern
- Preparing a table with rows for each person or group of persons concerned and columns for the four ethical principles: autonomy, non-maleficence, beneficence, and justice

Let us start this chapter with an example of a concrete ethical dilemma.

Marco, aged 3, was seriously hurt in a traffic accident. Upon arrival to the intensive-care unit, he is in severe hemorrhagic shock. His father, a Jehovah's Witness, prohibits blood transfusion, even after being told that immediate blood transfusion, followed by an operation is the only possibility for survival. The physician has to decide whether he will order blood transfusion or follow the father's demand.

The possible actions are blood transfusion against the father's will or no transfusion. The persons involved are Marco, his family, and the physician. In addition, we will also add to the list the community of Jehovah's Witnesses since they might be hurt if they learn that their religious belief was ignored. This brings us to the following table:

	Autonomy	Non-maleficence	Beneficence	Justice
Marco				
Marco's family				
Physician				
The community of Jehovah's Witnesses				

The physician's decision to apply blood transfusion against the father's will is to be compared to the alternative option—no transfusion, an option that would almost certainly lead to Marco's death. We will now insert ethical benefits, marked with a "+", and ethical costs, marked with a "−". Assessment of ethical benefits and costs is clearly subjective; however, subjectivity is an unavoidable characteristic of every ethical deliberation.

	Autonomy	Non-maleficence	Beneficence	Justice
Marco	?	+ + +	+ + +	+ +
Marco's family	− − −	+ + +	+ + +	+ +
Physician	+ + +	+ +	+ +	
The community of Jehovah's Witnesses	− −	+	+	+

As we can see, the ethical benefits of a transfusion against the father's demand greatly outweigh the ethical costs. The physician in the intensive-care unit has to make decisions without delay. If an agreement between doctors and parents cannot be obtained, some countries provide an urgent court authorization for assessments of lawfulness. If there is really no time, the doctor would institute the blood transfusion but seek retrospective court approval for his or her emergency life-saving treatment provided in the child's best interests. In many countries, however, no such urgent court service is available and physicians have to make a decision and accept full responsibility. In such a case, a physician rightly chooses a transfusion in order to give Marco a chance for survival but should not ignore the ethical costs of her/his decision. She will later speak with both parents, including with the mother who was not present at the time the boy was brought to the hospital. In a calm conversation, she will do her best to relieve the stress of the parents. She will assure them that she respects the very positive aspects of their religion. Furthermore, she will tell them that some Jehovah's Witnesses do not refuse blood transfusions and that their religion does not prohibit organ transplantation. Since blood transfusion is only the temporary transplantation of a liquid organ, the teaching against blood transfusion (but approving organ transplantation) is, in fact, not logical. The physician will also emphasize that the decision for blood transfusion is her decision, not theirs. All members of the intensive-care unit will be clearly reminded of the obligation to keep the transfusion confidential, so as to prevent leaking the story to the public or to journalists, which would bring an unnecessary burden to all involved. She might even "forget" to include information about the transfusion into the final medical report to be sent to the family physician. After all these measures, some minuses in the table may be reduced.

Let us now consider another case, similar at first glance, yet quite different after ethical analysis.

Damian, aged 12, has been treated for neuroblastoma for two years. Despite intensive treatment with three lines of cytotoxic drugs, he is now in relapse. Physicians agree that there is no indication for further active treatment and recommend only supportive and palliative treatment. Due to anemia, Damian is weak and dizzy. A blood transfusion might lead to short-term improvement of his symptoms but clearly cannot reverse the course of the disease. As Jehovah's Witnesses, his parents refuse blood transfusion.

The possible actions are a transfusion against the parents' will or no transfusion. The persons involved are Damian, his family, the physician, and the community of Jehovah's Witnesses. Here is the table, as prepared for the decision to order blood transfusion against the will of the parents.

	Autonomy	Non-maleficence	Beneficence	Justice
Damian	− −	?	+ (?)	?
Damian's family	− − −	− −	− −	− − −
Physician	+	?	?	?
The community of Jehovah's Witnesses	− − −	− − −	− −	− − −

As we can see, the transfusion brings very few benefits and severe ethical costs, which no lengthy explanation can reduce. As a teenager, Damian understands the religion in which he was raised. For the parents, the extremely painful experience of the imminent death of their child would be even worse due to a lack of respect from physician's side. Their sorrow and anger might spread to the whole community of Jehovah's Witnesses.

The ethical analysis may also be used for more general discussions. Let us use this approach in the assessment of the obligatory notification of all new cancer cases to a cancer registry.

On the basis of legislation on the obligatory registration of cancer cases, the Cancer Registry of Slovenia was established in 1950, with the task of analyzing data on the diagnostics and treatment of patients with cancer in Slovenia. Thus far, individual consent to collect and analyze personal data has not been required. Recently, an initiative has been presented to change the law. On the basis of the ethical principle of autonomy, the proponents insist that patients with cancer have the right to consent before being included in the database. Members of the Parliament do not understand that this would lead to the incomplete registration of patients with cancer and would severely undermine the professional and scientific role of the registry.

The persons involved are patients with cancer; all citizens as potential future cancer patients; public services in charge of control of the quality of medical care; public services in charge of the rational planning of medical facilities; and scientists (oncologists, ecologists). Let us prepare a table for ethical analysis, comparing a decision not to change the law against a proposal to require individual consent.

	Autonomy	Non-maleficence	Beneficence	Justice
Patients with cancer	− −	−	+	+ +
All citizens	+ + +	+	+ + +	+ + +
Quality control	+ + +	+	+ + +	+
Planning of healthcare facilities	+ + +	+	+ + +	+
Oncologists, ecologists	+ + +	+	+	?

The analysis shows ethical costs for patients with cancer in the fields of autonomy and non-maleficence. This indicates the importance of the stringent protection of personal data: personal information in a public database becomes problematic only in the case of its misuse. However, most other fields reveal ethical benefits. Analyses prepared by the registry offer information on the quality of cancer care, on the pressing needs for the better organization of cancer care, and on the connections between the incidence of cancer and environmental factors. These data benefit current and future cancer patients. Twenty years ago, data from the Cancer Registry convincingly demonstrated that nearly half of Slovenian patients with mesothelioma lived within 30 km of a large asbestos factory, leading to the halting of production of asbestos. Another example was an episode of great public concern, following the news of two cases of childhood leukemia near a nuclear power plant. The data from the registry were decisive in resolving the panic, showing that in over 20 years of the operation of the plant, the local incidence of leukemia and of other cancers was not above the national average.

Should we use similar arguments to support a proposal for a national registry of psychiatric patients? Most probably not, since we may expect greater ethical costs and fewer benefits. In comparison to patients with cancer, the stigmatization of psychiatric patients is much more pronounced, and eventual unauthorized access to the database would lead to severe ethical costs. In addition, no benefits may be expected from the prevention or early detection of psychiatric diseases.

What about a permanent registry of condemned pedophiles, without the possibility of being erased? There is no doubt that ethical analysis would reveal ethical costs in the fields of their autonomy, non-maleficence, and beneficence. Still, they should be aware of this possibility when they perform their revolting activities. While some will insist that pedophiles should have the right to start life anew, without the burden of the past, we should also consider those parents who do not wish such a person to become a coach for their children or a young widow with three children who will rent a room in her house to a stranger. It is therefore not logical to protect the autonomy of a pedophile and, at the same time, to limit the right to information and hence the autonomy of all others who have a legitimate interest to be informed about his past.

It is clear that ethical analysis involves both ethical and professional arguments. Ethical analysis is a strong tool against one-sided presentation of ethical dilemmas. Attention to a single dimension, most often to respect for individual autonomy, should not obscure the ethical benefits and costs for other persons whose interests are also at stake. It is clear that ethical benefits and costs are subject to individual interpretation. Ethics is not mathematics, and conclusions from an ethical analysis may be debatable, based on different professional arguments. Thus, ethical analysis does not provide a single answer, but rather a framework for balancing and reaching a rational conclusion.

Suggested Reading

Cottone RR, Claus RE. Ethical decision-making models: a review of the literature. J Couns Dev. 2000;78:275–83.

Davies R, Ives J, Dunn M. A systematic review of empirical bioethics methodologies. BMC Med Ethics. 2015;16:15. https://doi.org/10.1186/s12910-015-0010-3.

de Vries R, Gordijn B. Empirical ethics and its alleged meta-ethical fallacies. Bioethics. 2009;23:193–201. https://doi.org/10.1111/j.1467-8519.2009.01710.x.

Fowler MD. Ethical decision making in clinical practice. Nurs Clin North Am. 1989;24:955–65.

Kirklin D. Minding the gap between logic and intuition: an interpretative approach to ethical analysis. J Med Ethics. 2007;33:386–9.

Macauley RC. The analysis and resolution of ethical dilemmas. Handb Clin Neurol. 2013;118:11–23. https://doi.org/10.1016/B978-0-444-53501-6.00002-0.

Parker M. Two concepts of empirical ethics. Bioethics. 2009;23:202–13. https://doi.org/10.1111/j.1467-8519.2009.01708.x.

Taylor RM. Ethical principles and concepts in medicine. Handb Clin Neurol. 2013;118:1–9. https://doi.org/10.1016/B978-0-444-53501-6.00001-9.

Autonomy and Its Limitations

6

Abstract

As the leading ethical principle in the Western world, autonomy deserves a detailed discussion. Understood as the right to self-determination, autonomy includes the right to information and protection of privacy. In an ideal situation, a patient with full autonomy participates in all essential medical decisions and consents to every invasive procedure. However, even patients with full capacity have the right to transfer their autonomy to others: to a family member, friend, or to his/her physician. In cases of patients unable to decide for themselves and therefore with limited autonomy, surrogate decision-making is justified. Two aspects of surrogate decision-making deserve special attention. First, loss of autonomy is rarely complete: every person should be offered appropriate information and allowed to participate in decisions within her/his capacity. Second, surrogate decision-making should be based upon the ethical principle of beneficence and not upon autonomy. In other words: while a person with full autonomy may refuse a life-saving treatment, a physician is not ethically obliged to respect a directive by a surrogate decision-maker if this directive is clearly against the patient's interests. Finally, some persons make advance directives, to be followed in case of their future incapacity to participate in decisions regarding their treatment. While such written or oral directives are helpful, their validity may be reconsidered in situations that the person could not foresee at the time of making the advance directive.

Dr. Lewis Thomas was a physician, immunology researcher, dean, poet, etymologist and essayist. He was known worldwide for his columns in the New England Journal of Medicine, which later appeared in his book *The Lives of a Cell—Notes of a biology watcher* (The Viking Press, 1974). One of his key metaphors was a comparison of humans to ants. Many animals act as a large organism: a single ant cannot survive, and only an anthill possesses all the functions of an organism. Are humans much different? How strictly should we understand the autonomy of an individual?

Among the four ethical principles, respect for autonomy is of such importance that it deserves its own chapter. We usually speak about the respect for the autonomy of an individual, but the principle also applies to groups of persons, such as the respect for the autonomy of physicians or for university teachers. While our discussion focuses on medicine, we have to emphasize that the same rules are also valid when discussing the autonomy of an individual in another context: in relation to government, to media, or to professional associations. In these relations, politicians, opinion leaders, or chairpersons should recognize the right to autonomy of those who are on the weak side.

Let us begin with a brief reflection on patients' autonomy in the past. It is only since the second half of the twentieth century that patients' autonomy has been brought to a central position in our discussions on medical ethics. For centuries, very few spoke about patients' right to be informed and to participate in decisions regarding their treatment. The extent of information was left to physicians who also made decisions and carried on the treatments which they considered to be the most appropriate. It is interesting that such a clearly paternalistic attitude prevailed in the times when medicine was most often powerless, when diagnoses were more than uncertain, and when treatment methods were often ineffective, if not overtly harmful. In the early twenty-first century, we are in a different situation. Nowadays, most diagnostic and therapeutic methods are on a firm scientific basis. Despite this fact, we simply cannot imagine excluding a patient from the information and from the process of making a decision.

The word *autonomy* is derived from the Greek words *auto* (self) and *nomos* (law)—one who gives oneself one's own law. For full implementation of autonomy of an individual, the following conditions have to be respected:

- Right to information
- Right to confidence
- Right to privacy
- Ability to deliberate and formulate a decision
- Ability to implement a decision, and the right that the decision is respected

6.1 Right to Information

Full information is an essential basis for any rational process through which an individual can form an opinion and reach a decision. The duty to inform is, therefore, among the basic tasks of every physician. Due to the importance of this topic, the next chapter will be dedicated to communication and information in medicine.

Regarding language, form, and contents, information should be adapted to patient's understanding; it should contain all essential facts upon which a decision is built. With every chronic disease, presenting information to a patient should be regarded as a process, rather than a single event: information changes along with the dynamics of the disease and our deeper understanding of its course.

The patient's right to information corresponds to the physician's virtue of veracity.

6.2 Right to Confidence

In this brief overview, the right to confidence and the right to privacy are discussed within the broad principle of respect for autonomy. These two rights might well fit also under the ethical principle of non-maleficence.

Since the time of Hippocrates and other ancient healers, respect for confidence has been among the basic physician's duties. Disrespect of confidence may affect patients' autonomy since individuals not authorized by the patient may influence decision-making as well as infringing the patient's privacy often regarded as an aspect of autonomy. Other patients' rights and interests may also be affected. As an example, spreading the information about a genetic predisposition to a disease may affect individual's interests regarding education, employment, health and life insurance, or choice of partner, and may also affect the lives of other family members.

Commitment to confidence may be disregarded only in two situations: upon a patient's permission, or in the case of public interest, as specified by law and by an ethical code. A patient may authorize the physician to reveal information on his/her disease to relatives or others. From the ethical standpoint, oral permission is sufficient; however, lawyers often require a written document, usually for evidential reasons. In addition to such *explicit consent*, situations may arise when *implicit consent* to reveal patient's personal data was made. We speak about such implicit consent when a patient holds out an arm to have blood pressure taken. But it might also be relevant when a patient or his relatives widen the field of confidence and openly speak about the disease in public. Here are two such situations.

When talking to a journalist, 24-year-old J.K. told her that he had advanced testicular cancer with lung metastases. During chemotherapy at the Institute of Oncology, he suffered from severe nausea and vomiting and, therefore, declined further treatment. Instead, he contacted a healer who proposed goat milk and treated him with bio-energy. Since then, three years have passed. He feels healthy, and chest X-ray and tumor markers are now normal. A paper entitled "Natural treatment cured him of cancer" appeared in a local tabloid.

Is it acceptable for his oncologist to send a letter to the journal, explaining that the cure for this cancer is due to chemotherapy and not to the alternative approach? After five (of the planned six) cycles of intensive chemotherapy, the lung metastases disappeared, and tumor markers returned to normal values. May he assert that the statement that the "official medicine had thrown up its hands" is not true? Most lawyers would say that the doctor may reveal personal information only after the patient's explicit consent. From my own ethical standpoint, however, it should be not only acceptable but indeed highly justified to offer an explanation that would be beneficial to other patients who might follow such distorted reports.

Here is another example of implicit consent.

In an interview published in a national weekly magazine, the father of 12-year-old Domen accused a physician of committing malpractice. At the age of 3, the boy was brought to a hospital due to pyrexia. The diagnosis of meningitis was made only three weeks later. Due to a delay in diagnosis, Domen is severely handicapped and is mentally on the level of a 2-year-old child. According to his father, the doctor should be prosecuted for the consequences of malpractice. The name of the accused physician appeared in bold letters on the cover page of the magazine.

After such a public attack, should the physician be allowed to defend herself by revealing the whole story? May she tell that the course of disease was very atypical for meningitis? The first two lumbar punctures were negative, and only the third lumbar puncture confirmed meningeal inflammation. The diagnostics were performed according to all standards, and there is nothing to support the accusation of a professional malpractice.

In ethical consideration (but not necessarily so in the legal perspective), it is the patient who defines the field of confidence. In case of a complaint against his physician and addressed to the Medical Chamber, members of the Ethical Committee are informed about his complaint and seek an explanation from the physician. Still, they are bound by the rule of confidence and therefore should not reveal the story to anyone outside of their circle. However, if the patient or his/her relatives bring the issue to the public, then I believe that the physician has every right to address the same audience with his explanation.

Finally, we have to add a comment about breaking confidence due to a higher interest—the interest of other persons, or of society at large.

Airplane pilot Andreas Lubitz, aged 28, attended a psychiatric out-patient clinic. Due to Lubitz's tendency towards suicide, the psychiatrist declared him to be unable to work. Lubitz kept this report to himself and went back to work as a co-pilot. On March 24, 2015, when flying over the Pyrenees, he locked himself into the flight deck and intentionally crashed the plane, killing 158 people.

I firmly believe that the psychiatrist should have sent his report to his employer, the airline: we are dealing with a higher public interest which should have priority over the patient's right to confidentiality. Even if the chances of committing suicide were slight, the consequences were so grave that the airline company should have been contacted.

In ethical reasoning, the Lubitz case is relatively simple. The physician's breaking the obligation of confidentiality would remain undisclosed if Lubitz had kept his promise and remained on sick leave, instead of returning to work. In contrast, the tragic consequences of non-compliance with the physician's advice are extreme. More delicate is a situation with a psychiatric patient who would tell his physician about his plans to harm somebody from a narrow circle, such as a spouse or other family members. In a dilemma between professional confidentiality and responsibility for those mentioned by the patient, the physician knows that warning the persons in question, or even a request for protection by police does not guarantee definitive protection. At the same time, the patient will notice the protective measures and will not trust the physician in their future relationship.

6.3 Right to Privacy

While the obligation to confidentiality corresponds to the right of a patient to control the flow of personal information, respecting privacy should protect a patient's right not to be disturbed in his private world.

The degree to which the right to privacy is respected may be the most obvious difference between state-owned and private providers of healthcare. Inadequate conditions in overcrowded hospitals lead to frequent disrespect of the right to privacy. During hospital rounds, patients are often interviewed or examined in the presence of other patients in the same room. However, we should not blame only the old hospital facilities: quite often, physicians and other health personnel do not even notice that their conduct is inappropriate. In the next chapter on communication, we will emphasize that a patient should be regarded as a person, rather than merely a bearer of a certain diagnosis. While the conversation with a patient normally includes some non-medical issues, we have to be aware that only a thin line separates the questions addressed to reveal patient's personal life from those that may be considered as an intrusion into patient's privacy.

In activities related to teaching and research, the protection of patients' privacy is of special importance. A request for a patient's permission to present her to a group of students should be taken seriously, and her eventual non-cooperation should be accepted as normal. In medical research and especially in designing personal questionnaires, the topics related to patients' privacy should be included only if absolutely necessary for the purpose of the research, with an option to omit these answers.

6.4   Voluntary Surrender of Autonomy

A person who is adequately competent, able to understand the situation, and able to decide may also waive his right to autonomy. In medical practice, this is by no means rare. "Doctor, you know, please do what you think is best for me,"—all physicians have often heard such words.

In such a situation, the physician should nevertheless explain all circumstances. Continuous information on the plan for diagnostics and treatment also applies to a patient who has transferred his autonomy to a physician. In case of a new situation, such as a plan for a new treatment, the physician should again offer the patient the opportunity to express his or her preference regarding treatment options.

6.5   Patients' Autonomy and Cultural Diversity

In the world of great cultural, religious, social, and economic diversity, significant differences are present regarding attitudes towards individual autonomy. The European and North American view on individual autonomy cannot be simply applied to other parts of the world. As we move to the South and East, we can see a marked imbalance of the social role and rights of women in comparison to men. In addition, the interests of the extended family, religious community, or the state often prevail over the interests of an individual. This is also reflected in communication and in decision-making in which the will of an individual and especially of a woman is of lesser importance.

I have no personal experience regarding the situation in which a physician with the European cultural background moves to Africa, the Middle East, or Asia, and I, therefore, cannot offer clear advice. It seems that the physician will have to comply with the local norms and relations. At least to some degree, the physician's attitude towards patients, family members, and wider society will be adapted to their expectations. However, a different situation arises when a physician cares for a patient who immigrated from different cultural surroundings. In my view, in such cases, it is the immigrant who has to adapt. While much effort and time are needed for communication and for reaching a decision, there should be no dilemma: in our cultural and ethical tradition, the individual and his or her interests and preferences have priority, regardless of whether we are dealing with a patient of our nationality or with an immigrant.

6.6   Persons with Limited Autonomy

In the realization of individual autonomy, the key elements are a critical reflection of the personal situation, understanding of the information, the ability to formulate a judgment, and the ability to decide. When these requirements are not met, we speak about persons with limited autonomy. This category includes:

- Children
- Unconscious patients and patients with limited consciousness
- Patients with severe emotional stress due to unexpected serious illness
- Patients with severe psychiatric diseases, here including mental retardation and senile dementia

It is of critical importance to understand that the dilemma between adequate and limited autonomy is not a question of all-or-nothing. Patients with limited autonomy are still able to decide about some issues regarding their way of life and treatment. They should be allowed to decide about all options that remain within their ability to understand. Even if a patient is clearly not able to make a rational decision, the physician should offer information on the disease and treatment, adapted to his or her level of understanding. Limited autonomy is therefore not a reason to abandon the duty of informing the patient.

In the past, it was the physician who decided on issues that were beyond the ability of a patient with limited autonomy. Even today, such a paternalistic approach is often the only real option in emergencies, such as immediate assistance to an unconscious patient when relatives are not present. In all other circumstances, support and agreement for a medical decision should be sought from the patient's surrogate decision-makers or on the basis of the patient's advanced directives.

6.7 Surrogate Decision-Makers

We speak about surrogate decision-makers when another person, the guardian, decides in the name of a person unable to make a decision herself. Parents for children and close relatives for others are usually those who accept the burden and responsibility for a decision. Occasionally, a guardian would present a written authorization: either signed in advance by the patient, or by the court in the case of mentally incompetent patients. More often, the guardian is a family member who does not have a formal document to authorize his role, in which case we have to trust and take him at his word. In such a case, some caution is recommended, especially when dealing with more complicated family relations such as divorced and remarried persons. When the issues of finances or inheritance are at stake, the physician should strive to reach a consensus of a wider circle of family members.

Respecting the ethical principle of autonomy is the most important, and indeed the only valid basis when a person with full autonomy decides about herself. Clearly, a person with full autonomy may insist on a decision that is against her best interests, and a physician has to respect such a decision. In cases of surrogate decision-making, the situation is different. Surrogate decision-making does not refer to the ethical principle of autonomy, but rather to the ethical principle of beneficence. When making a decision about a medical procedure for my child or for my elderly parents with dementia, I should not defend a decision that would be harmful to them. Children are not parents' property, and parents share the duty to act in their best interests. The physician has the right and the duty to defend a child or another person with limited autonomy when a surrogate decision-maker acts against the patient's best interests. In cases of disagreement, the interests of the child, or of another person with limited autonomy should prevail. The most appropriate resolution of such a disagreement is a rapid court decision. However in many countries including my own, Slovenia relying on a court decision would bring an unacceptable delay. For situations when a decision has to be made by physicians themselves, Article 16 of the Slovenian Code of Medical Ethics states:

> If the advance directive of a patient who is not capable of decision-making with respect to his or her own care is not known, the physician shall, in consultation with the patient's close relatives, propose treatment that is most beneficial for the patient according to the physician's judgment. If the consultation procedure results in disagreement, a medical council shall decide on the patient's greatest benefit and shall inform the patient's close relatives about the decision. If the disagreement is not resolved at that point, the physician is not bound to comply with the relatives' divergent opinion.

6.8   Advanced Directives

In cases of limited autonomy, the patient's advanced directive is the second possibility to support a medical decision. Advanced directives are a logical consequence of the clash between the widening options of modern medical technologies to prolong life, and the increasing awareness of the right to control one's own life, here including dignity when approaching the end of life. Elderly, or patients with severe chronic diseases often wish to express their preference regarding treatment in case of future incompetence and deterioration of their health. Since nobody can envision all possible future scenarios, the contents of such advance directives are usually quite broad and include statements such as refusal to be admitted to an intensive-care unit, or refusal of resuscitation. While advanced written directives remain rare in many countries, it is clear that this practice will become more common in future.

Advanced directives may be written and signed, sometimes in the presence of a family physician. They may also be expressed orally, preferably in the presence of at least two witnesses.

An advanced directive has no absolute validity. First, there is a time limit: many laws and regulations limit the validity of patients' advanced directives to 5 years, after which the person should renew its validity. In my view, a physician should always strive to understand whether the patient really envisioned the current situation at the time when formulating an advanced directive. The nature of the current medical problem is important: while a successful antibiotic treatment of acute pneumonia may quickly return the patient to her previous state, such a scenario is clearly different from a decision to start long-lasting artificial ventilation. Even when presented with a clear advanced directive, the physician should speak to a frail patient and try to get his or her current preference: treatment including intensive medical procedures or limited to palliative care. The physician should be aware that signing a sentence stating "Please let me die" when in relatively good health and for some abstract future time is much easier in comparison to reaching such a decision when the time comes.

In situations at the end of life, a patient's advanced directive is of precious help to a physician but has no absolute validity.

6.9 Physician's Autonomy and Conscientious Objection

Many important medical decisions are formulated in a group, such as a team meeting about newly diagnosed patients. When all relevant aspects are considered, and a balanced decision is reached, such a process is clearly beneficial to the patient. In collective decisions, the opinions of senior and experienced physicians usually prevail, and it is rare that a young physician would be successful in advocating an alternative option.

The next step is the implementation of the decision made by a team. Here, we are dealing with a personal rather than collective responsibility. This leads us to a conclusion: a physician who disagrees with a decision of a medical team and thinks that the proposed action is not in the patient's best interest has the right and duty to decline to perform the procedure. In such a case, we are dealing with the physician's right to autonomy.

The second example of respecting physician's autonomy is conscientious objection. In this case, the physician declares, in advance, that a certain act is not compatible with his or her personal values. The physician should be aware of the fact that his conscientious objection may conflict with his regular professional obligations. Let us be concrete: a physician who opted for specialization in gynecology and obstetrics may, according to my personal opinion, still declare conscientious objection against performing abortions, especially if seeking employment in an institution where abortions are not performed. In contrast, it is difficult to support a gynecologist in primary care who declares conscientious objection against prescribing contraceptives. Such a physician should understand that control of human reproduction is not only among the basic human rights but is widely acceptable also for the majority of religious persons.

Suggested Reading

Chamsi-Pasha H, Albar MA. Doctor-patient relationship. Islamic perspective. Saudi Med J. 2016;37:121–6. https://doi.org/10.15537/smj.2016.2.13602.

Coleman AM. Physician attitudes toward advanced directives: a literature review of variables impacting on physicians attitude toward advance directives. Am J Hosp Palliat Care. 2013;30:696–706. https://doi.org/10.1177/1049909112464544.

Ho A, Spencer M, McGuire M. When frail individuals or their families request nonindicated interventions: usefulness of the four-box ethical approach. J Am Geriatr Soc. 2015;63:1674–8. https://doi.org/10.1111/jgs.13531.

Quante M. Personal identity as a principle of biomedical ethics. Berlin: Springer; 2017. ISBN-13: 978-3319568676.

Robertson L. Contemporary interpretation of informed consent: autonomy and paternalism. Br J Hosp Med (Lond). 2016;77:358–61.

Schramme T, editor. New perspectives on paternalism and health care. Berlin: Springer; 2015. ISBN-13: 978-3319179599.

Sarafis P, Tsounis A, Malliarou M, Lahana E. Disclosing the truth: a dilemma between instilling hope and respecting patient autonomy in everyday clinical practice. Glob J Health Sci. 2013;6:128–37. https://doi.org/10.5539/gjhs.v6n2p128.

Tunney RJ, Ziegler FV. Toward a psychology of surrogate decision making. Perspect Psychol Sci. 2015;10:880–5. https://doi.org/10.1177/1745691615598508.

Communication

7

Abstract

In medical education, communication is given far less attention, in comparison to diagnostics or treatment. Proper communication is of paramount importance, and weak communication is at the roots of most complaints or accusations of professional misbehavior. Communication should always be bidirectional. When dealing with relatively simple and transient medical problems, communication is often limited to technical instructions. With chronic diseases, communication includes much deeper insight into patient's personality, family, and social background, and should be understood as a process, rather than as a single encounter. Honest information about the results of diagnostic procedures and understandable presentation of the proposed treatment are essential. In contrast, the prognosis of a disease often remains uncertain and should be communicated with caution: statistical data have limited validity for an individual, and unexpected favorable or unfavorable course of a disease is not rare. In dealing with a serious disease, an experienced and compassionate physician understands that positive motivation of a patient and of his/her family or supporting team is of crucial importance.

© Springer Nature Switzerland AG 2019
M. Zwitter, *Medical Ethics in Clinical Practice*,
https://doi.org/10.1007/978-3-030-00719-5_7

Diagnostics, treatment, and communication as the three pillars of medicine may be compared to a three-legged stool: if one leg breaks, the chair collapses

The three pillars of medicine are diagnostics, treatment, and communication. These may be compared to a three-legged chair: if one leg breaks, the chair collapses. Everyone will agree that we cannot choose appropriate treatment without having established an accurate diagnosis. Similarly, the importance of selecting the best possible treatment is obvious. However, despite accurate diagnostics and the best treatment, physicians can fail if they are not able to communicate their knowledge to the patient in a clear and understandable manner and if they do not adjust their actions to the values and expectations of the patients. Inappropriate and poor communication is at the roots in most allegations of malpractice.

7.1   Information Flows in Multiple Directions

In communication, information flows from the patient to the physician and of course in the reverse direction, from the physician to the patient. Only in this way can they jointly establish the diagnosis and choose treatment options. Frequently, they will also involve other persons in the process of communication, for example, medical nurses, social workers, and occasionally a psychologist. On the patient's side, family members, friends, and co-workers may participate in the communication. Due to the requirement for multidirectional communication, information should not be limited to written explanations. Booklets, informative leaflets, and other forms of written explanation are certainly great support materials; however, these cannot replace personal conversation.

7.2 Communication as a Process

In simple medical interventions, such as minor injuries, the focus of communication is on technical instructions. The situation is different when dealing with severe and/or chronic diseases.

Tamara, aged 31, a never-smoker and mother of two children, worked as a financial consultant in a research institute. Due to persistent back pain, a radiological examination was done, leading to a temporary diagnosis of a lytic lesion in a pelvic bone. Further diagnostics discovered a primary tumor in the right lung, and biopsy confirmed the diagnosis of adenocarcinoma. With positron emission tomography, multiple bone and liver metastases were clearly visible, thus confirming the advanced stage of the disease. As the initial treatment, palliative radiotherapy to the painful bone was proposed; however, the plan changed following an additional pathology report: the tumor was positive for mutations of epidermal growth factor receptor. She consented to participate in a clinical trial of intercalated chemotherapy and a tyrosine kinase inhibitor. This treatment led to a complete response: she returned to work and enjoyed life with her family. Two years later, however, multiple brain metastases appeared. Palliative radiotherapy had only a temporary effect, and three years after the initial diagnosis, she died in the palliative unit of a local hospital.

At every step of this case, proper communication was an essential task of the physician. During the 3 years of her disease, 11 situations—and possibly more—may be identified when a detailed explanation of indications for diagnostics and proposals for treatment were needed and when her attitude, consent, or alternative opinion needed to be heard and considered. Gradual understanding of the characteristics of the disease, evolution of the disease process, and the patient's attitude towards the disease and treatment all changed with time and influenced communication.

Communication is a process rather than a single event. The second, equally important reason for arranging several meetings is that the patient and his relatives cannot absorb and understand extensive information about the nature of the disease, intended diagnostics, treatment, and supportive therapy during a single visit. We, therefore, must schedule several meetings and make it possible for them to ask questions about whatever might have remained unclear from previous visits.

7.3 Modern Media and the Internet

Many physicians feel irritated when admitting patients who seek opinion on information and advice they found on the internet. The physician often has the impression of being a part of a communication triangle: the physician, the patient and his relatives, and an anonymous adviser from the web. Should the physician allow that the process of decision-making involves someone who does not know the patient in question, someone who is not aware of truly feasible diagnostics or treatment options and who obviously takes no responsibility for his or her advice? How should the physician respond to information that does not rest on the scientific foundations of contemporary medicine and is evidently driven solely by financial interests?

A bad mood, feelings of superiority, or outright rejection of all sources of information from the media or the web can only lead to conflict. It is true that internet-driven information overload requires clarification, which takes the physician's precious time, often exceeding the limited minutes planned for the patient's visit. Nevertheless, the physician should maintain positive attitudes towards the information available on the web. Physicians must understand that this is not a sign of mistrust but rather an attempt by the patient and the relatives to find hope despite the perhaps distressing prognosis. At the same time, such scenarios remind us that physicians should be actively involved in curating objective information on the web. The physician must understand that denial is not productive. The internet is a reality, and providing objective information is the only solution for alleviating the problem of biased information on the web.

7.4   The Broad Scope of the Conversation

The contents of the conversation between the physician and the patient are of course strongly determined by the nature of the medical problem. In communication about diseases that have a major impact on the patient's life, it is crucial that the physician and the patient establish a good interpersonal relationship during the very first encounter. When I meet a new patient suffering from lung cancer, I simply cannot start the conversation by asking: "Are you coughing up blood?" The patient is often in distress, deeply worried, and uncertain. I can lead him out of this state by asking where he came from and what he did for a living. If it turns out along the way that I happen to know the patient's village and his veterinarian, we have overcome the initial mistrust. This will allow me to adapt future conversations to the patient and his way of understanding. Even for me as a physician, the work is significantly more interesting if I treat my patients as fellow humans and not solely as carriers of diagnosis. This conclusion is borne out in the numerous records made by physicians; one can read countless surprising, often entertaining stories. What all these recollections have in common, however, is the link between the medical problem and the patient's life story. I will never forget a patient from Komenda, a small town near Ljubljana. I was asking him about his profession as a blacksmith. When we were discussing horseshoeing, I asked (being an urban dweller who never lived in the countryside) if he was scared when nailing the hooves. "What if the horse kicks?" "No, doctor. First, you approach the horse, stroke his mane and talk to him; then he will lift the foot up by himself." The same holds for patients: kindness during the first encounter opens the door to good cooperation.

The conversation about the disease interweaves the topics of diagnosis, treatment options, disease prognosis, and patient motivation. Frequently, the disease is recognized only gradually, and this must be reflected in communication. There is nothing wrong with telling the patient that I do not yet know everything. Acknowledging such uncertainty is even more important when giving the prognosis.

The experienced physician knows that statistical predictions cannot be projected onto the individual patient and that the course of the disease can always foster exceptions, for better or for worse. The sentence "Doctors gave me six months to live" may be heard in clichéd TV shows but has no place in the physician's office. Fortunately, physicians in Europe are not yet in a situation as in the USA, where a physician was sued for not having presented the precise survival statistics to the patient. In my approach, I prefer that of Srečko Katanec, former manager of the Slovenian soccer team who knew how to motivate his soccer team and led them to victory against objectively stronger opponents, world champions from Italy. Indeed, precise statistical predictions may be very discouraging. Every patient has the right to hope for a miracle, or at least to hope to be on the better side of the Gaussian survival curve. Every patient represents a unique narrative, and one of the physician's important roles is to motivate the patient and to not shatter his hopes.

When upon learning about the serious disease, the patient experiences distress or crises—the best help we can offer is to motivate him to be active. The physician informs the patient about his options for improving his chances: instructions for physical activity, healthy diet, good hydration, and breathing exercises. Some patients even believe homeopaths, other sorts of wizards, and their "bioenergy." Although these therapies are certainly driven by mere placebo effects, they can offer psychological support to the patients as additional measures. Of course, the basic treatment must not be discontinued in such cases.

7.5   The Opaqueness of the Medical Jargon

A great obstacle to communication is the medical language that is often incomprehensible to persons with no medical knowledge. Many among us have been asked by friends to "translate" a report of medical examination or a hospital report into more comprehensible language. It is, of course, understandable that we use professional terminology in written documents; however, in a conversation with the patient, we need to assess whether he/she can understand us. Someone without medical education will have a difficult time understanding what we mean by "adjuvant therapy." When I see a woman who, despite a successful cancer operation, needs additional (adjuvant) therapy, I explain it as follows: "Imagine we cleared the weed from your garden by cutting it with shears. We might not see the weed right now; yet, if this is all we do, it will grow back." When explaining the diagnosis and proposals for treatment, analogies with everyday life can be of great help.

Should we avoid using the word "cancer" when talking to the patient? Should we rather use a different word such as "lump," "thickening," "tumor" instead? No, when deciding about treatment options, we cannot avoid using the word "cancer." Once I get to know the patient better, and I refer her to control examination after treatment, I can relax the conversation somewhat and say that "We are simply trying to see whether the tenant has left already." At the right time and in the right amounts, an entertaining tone can fit the conversation well; we must not, however, give jokes as answers to serious questions—the patients might rightly think that we are making fun of them.

We all know physicians who see conversations with patients as a source of annoyance. Most often they prefer to limit themselves to technical instructions: "Please come tomorrow at 11 AM to…". These physicians will most often say that they have no time for conversations with patients. The excuse about the lack of time simply does not hold. We have already seen that poor communication is the root cause of most allegations of malpractice. In the case of public allegations, the same physician will lose incomparably more time when trying to defend his actions in the media and in court.

7.6   Cultural Diversity

Cultural diversity is growing in society at large, and clearly also in medicine. In patient management, we cannot disregard the patient's system of values and the specific environment from which the patient comes. Our sensitivity to cultural diversity and our ability to adapt our attitudes towards patients accordingly is one of the fundamental requirements for good communication.

In communication with patients, there is no place for stereotypes or nationalism. Patients coming from underprivileged communities need closer attention because it is more difficult for them to claim their rights by themselves, and they have more difficulties expressing their wishes and expectations. While we understand that many foreign patients do not speak our language, some have difficulties accepting the fact that certain immigrants are not fluent in the language of their host country. Yet, between the ghetto where they live and the construction site where they work, there is neither the desire nor the opportunity to learn what for them is a foreign language. The physician must realize that the contents of the communication are not determined by what the physician said but by what the patient understood.

7.7   Practical Advice

Often though of course not always, both husbands and wives find it helpful if they can enter the physician's office together, as a pair. Their joint understanding will be greater, and there will be fewer misunderstandings about what the physician and the nurse told them. It will also be easier for them to continue the conversation at home.

We often see that disease brings couples closer together. Old resentments are forgotten and together they fill their relationship with new content. Care for an ill partner can even become excessive. Especially when a wife takes care of her ill husband, we sometimes see a disturbing insistence on feeding, an excessive prohibition of even a single drop of alcohol or a much craved for cigarette, and disproportionate attention to everything, even very minor signs of disease.

In terms of communication, the most difficult group in severe or chronic disease treatment is single men. They find the new situation difficult when suddenly they are not completely independent anymore and must rely on help from others. These men have hard times acknowledging their helplessness. Some of them may be victims of the society of "permanently adolescent" boys that works perfectly in a pub, but quickly forgets those who no longer sit at the bar. The situation is different for single women who often build a circle of colleagues and friends whom they can ask for help when in distress.

Let us conclude this chapter with practical advice for those readers who feel and know that medicine is more than just a technical routine. Communication requires readiness, space, and time. Even if the physician is sincerely willing to have a conversation with the patient, it is undoubtedly true that he is not available all the time for a long conversation. In my experience, it is extremely important that the physician schedules in his agenda office hours, days and times when he is available to talk with patients and their relatives. Otherwise, patients and relatives will be chasing him in the office, the ward, in breaks during surgeries, during preparation for a difficult procedure or on a day off after being on duty. When relatives stop the physician in the hallway right before a meeting or a lecture, they often incorrectly assume that he knows all the patient data and examination results. Such a passing and unresolved conversation can be very counterproductive. Publicly announced office hours are not a waste of time, but instead make for a better use of it.

Suggested Reading

Baider L, Surbone A. Cancer and the family: the silent words of truth. J Clin Oncol. 2010;28:1269–72. https://doi.org/10.1200/JCO.2009.25.1223.

Cocanour CS. Informed consent-it's more than a signature on a piece of paper. Am J Surg. 2017;214:993 7. https://doi.org/10.1016/j.amjsurg.2017.09.015.

Mallia P. The nature of the doctor-patient relationship: health care principles through the phenomenology of relationships with patients. Berlin: Springer; 2012. ISBN-13: 978-9400749382.

Marini MG. Narrative medicine: bridging the gap between evidence-based care and medical humanities. Berlin: Springer; 2016. ISBN: 978-3-319-22089-5.

Perez B, Knych SA, Weaver SJ, Liberman A, Abel EM, Oetjen D, Wan TT. Understanding the barriers to physician error reporting and disclosure: a systemic approach to a systemic problem. J Patient Saf. 2014;10:45–51. https://doi.org/10.1097/PTS.0b013e31829e4b68.

Pugliese OT, Solari JL, Ferreres AR. The extent of surgical patients' understanding. World J Surg. 2014;38:1605–9. https://doi.org/10.1007/s00268-014-2561-8.

Sharif T, Bugo J. The anthropological approach challenges the conventional approach to bioethical dilemmas: a Kenyan Maasai perspective. Afr Health Sci. 2015;15:628–33. https://doi.org/10.4314/ahs.v15i2.41.

Surbone A. Cultural aspects of communication in cancer care. Recent Results Cancer Res. 2006;168:91.

Surbone A. Truthfulness of more optimistic vs less optimistic messages for patients with advanced Cancer. JAMA Oncol. 2015;1:687–8. https://doi.org/10.1001/jamaoncol.2015.1166.

Surbone A, Baider L. The spiritual dimension of cancer care. Crit Rev Oncol Hematol. 2010;73:228–35. https://doi.org/10.1016/j.critrevonc.2009.03.011.

Surbone A, Baider L. Are oncologists accountable only to patients or also to their families? An international perspective. Am Soc Clin Oncol Educ Book. 2012:e15–9. https://doi.org/10.14694/EdBook_AM.2012.32.e15.

Surbone A, Zwitter M, editors. Communication with the cancer patient. Information and truth. Ann NY Acad Sci, Vol. 809; 1997. ISBN: 0-89766-985-1.

Surbone A, Zwitter M, Rajer M, Stiefel R, editors. New challenges in commnication with cancer patients. Berlin: Springer; 2013. ISBN: 978-1-4614-3368-2.

Zaner RM. A critical examination of ethics in health care and biomedical research: voices and visions. Berlin: Springer; 2015. ISBN-13: 978-3319183312.

Zill JM, Christalle E, Müller E, Härter M, Dirmaier J, Scholl I. Measurement of physician-patient communication—a systematic review. PLoS One. 2014;9:e112637. https://doi.org/10.1371/journal.pone.0112637.

Relations in the Medical Team

Abstract

Nowadays, medicine is almost always a team activity. While physicians readily ask for the help of colleagues of other specialities, some physicians are reluctant to recognize the expertise of other health professionals such as nurses. However, in the light of rapid expansion of knowledge in all fields of medicine, sharing of knowledge, expertise, and responsibility is essential. Within a healthcare team, discordances in professional and organizational issues are inevitable and should be resolved in an open discussion based on arguments, rather than on power. Authoritarian leadership is inappropriate as it leads to frustration, suppression of new ideas, and concealing of professional misconduct.

A boat trip for a positive team spirit

In the previous chapter, we focused on communication with the patient and relatives. We are left with the important question of relations within the medical team.

Until recently, the physician was considered to be not only the first but also the only one to decide whatever is at stake in his healthcare institution and the healthcare system generally. Nowadays, teamwork is a reality. Let us emphasize, without any pretense, that the patient is the healthcare team's sole focus of attention. Anything concerning interpersonal relations among the physicians themselves and between physicians and other professional staff members must be centered on the interest of the patient.

8.1   Professional Competence

The amount of knowledge about the causes of diseases, disease characteristics, diagnostic procedures, and treatment is incredibly large nowadays. It is therefore natural for physicians to turn to specialists for answers that extend beyond their own speciality. It is much harder to understand, however, why many physicians adamantly refuse to recognize that their own knowledge can be usefully complemented by the expertise of other domains, for example, nursing. We are only gradually incorporating these attitudes into our own mindsets. Because we physicians too often do not show enough concern for real teamwork and collaborative efforts with experts from other healthcare professions, they too (medical nurses, laboratory and radiology engineers, biochemists, etc.) all too often prefer to stay within their own professional circles. In some places, we can therefore already see two separate systems of patient management in place: medical and nursing treatments. It should go without saying that this is of no benefit to the patient.

8.2   Communication in a Medical Team

The same principles that we described in the chapter on communication with patients apply to communication in a medical team. Here, too, communication must be bidirectional, and it must be an ongoing process.

Bidirectional communication is essential in everyday work. Good communication with a medical nurse can help the physician in assessing the patient's condition, in forming a plan of treatment, and in conversations with the relatives. The physician is aware that he spends considerably less time with every individual patient than the medical nurse does; her assessment can, therefore, be immensely helpful. When faced with specific expert questions—for example on the treatment of constipation or in preventing and treating bedsores—he will consult the nurse for advice. Even in cases of decisions pertaining to invasive diagnostics in elderly patients whose health condition is poor, a young physician might take into account the advice of an experienced medical nurse who will be in a better position to judge that the patient is at the end of life. Clearly, the influence is bidirectional: the physician will explain to the entire team that this is an exceptional case and that despite the presently critical and—at first glance—hopeless situation, the team must do every-thing to follow slim chances for survival of the patient.

Bidirectional communication is especially necessary when introducing innova-tion in diagnostics and healthcare. No physician and no medical nurse should intro-duce a new approach before presenting it to the entire medical team.

When we discussed communication with patients and relatives, we emphasized the importance of good organization and office hours. It is similar for communica-tion in the medical team. Questions about relations, work organization, or profes-sional issues cannot be resolved "at the door." The physician should prepare a regular schedule of medical team meetings where professional and organizational issues can be discussed. Everyone should have the opportunity to put forward his or her proposal at the meeting before conclusions and decisions are formed.

8.3   Disagreement and Conflicts

Disagreements in medical teams are commonplace. We can distinguish between creative and destructive disagreements. When a disagreement arises, it is in the large majority of cases a creative disagreement: a team member sees the solution to a given professional or organizational problem differently than most other mem-bers or the team leader. A good team leader will be receptive to new ideas and changes and will initiate an open discussion of proposals. Unfortunately, many of those in leading positions see any proposal for change initiated by other team members as a threat to their authority. Such self-centered leadership will bring creative disagreement to a destructive phase. Wholly professional and organiza-tional issues become intertwined with personal resentments and disqualifications. This is to the detriment of the entire team and indirectly, obviously, the patient. Disagreements that have entered the destructive phase cannot be solved at the same level, which means they must be dealt with at the higher levels of management.

The relationships with patients and their relatives are similarly not always har-monious; tensions and open disputes can also arise there. Addressing the issues as soon as possible, honestly, without preconceived judgments or disqualifications, and with a sense of solidarity among all healthcare professions—these are the basic principles for solving the conflicts. It would be very inappropriate if the physician stepped back because, say, the conflict involves the patient, the relatives, and the medical nurse, as if the dispute were not the physician's problem. The physician leading a department or a clinic must accept the burden of his responsibility. On the one hand, he should appease the patients or relatives by offering an appropriate explanation, even an apology. He can thus prevent the dispute from entering a destructive phase which could potentially involve the media and lawyers. On the other hand, he should hold an open discussion with the medical team to analyze the origins of the dispute. When a leader assumes his responsibility in managing con-flicts with patients and relatives, this gives him the right to point out, amicably, inappropriate measures.

8.4   Positive Team Spirit

Let us conclude with a thought on positive team spirit. We spend a major part of our lives at work. It undoubtedly matters if we are coming to work relaxed and happy or embittered in unwilling. It is not just about personal feelings, but also about efficient, professional and patient-friendly service. We are talking about mutual help when we do not want our colleague to say "This is not my duty" and walk away. In the previous chapter, we have seen how spending a couple of minutes discussing non-medical questions with the patient can significantly facilitate communication. In relationships with our co-workers, a short chat on family-related questions or a team-wide afternoon retreat for bowling will help instill a positive spirit in the entire team.

Finally, I do not want to leave the readers with the impression that I assigned too many professional duties to the physician. Disappointments, even over actions taken by our professional peers, are unavoidable. In such cases, the family remains our only source of support. The physician should take care of his family and maintain harmonious family life when his professional environment is still in good order.

Suggested Reading

Churchill LR, King NM, Cross AW, editors. The physician as captain of the ship: a critical reappraisal. Dordrecht: D Reidel; 1998. ISBN-13: 978-9401737364.

Clausen C, Cummins K, Dionne K. Educational interventions to enhance competencies for interprofessional collaboration among nurse and physician managers: an integrative review. J Interprof Care. 2017;31:685–95. https://doi.org/10.1080/13561820.2017.1347153.

Kossaify A, Rasputin B, Lahoud JC. The function of a medical director in healthcare institutions: a master or a servant. Health Serv Insights. 2013;6:105–10. https://doi.org/10.4137/HSI.S13000.

Martínez-González NA, Rosemann T, Tandjung R, Djalali S. The effect of physician-nurse substitution in primary care in chronic diseases: a systematic review. Swiss Med Wkly. 2015;145:w14031. https://doi.org/10.4414/smw.2015.14031.

Niezen MG, Mathijssen JJ. Reframing professional boundaries in healthcare: a systematic review of facilitators and barriers to task reallocation from the domain of medicine to the nursing domain. Health Policy. 2014;117:151–69. https://doi.org/10.1016/j.healthpol.2014.04.016.

ter Maten-Speksnijder A, Grypdonck M, Pool A, Meurs P, van Staa AL. A literature review of the Dutch debate on the nurse practitioner role: efficiency vs. professional development. Int Nurs Rev. 2014;61:44–54. https://doi.org/10.1111/inr.12071.

Professional Malpractice

9

There are not many bad people. Working conditions are bad, not people.

Andrej Robida

Abstract

Five key issues on medical malpractice will be discussed.

1. Many cases of malpractice are settled with good communication and an excuse, when appropriate. In contrast, weak communication is at the roots of most cases that are brought to the attention of a wider community, to media, or to the legal system.
2. Accusations of medical malpractice often originate from misunderstanding of medicine as an accurate scientific discipline. Patients, their relatives, lawyers, and journalists do not understand that medical decisions are often based on incomplete or non-specific diagnostic procedures, and that the most probable diagnosis in a certain situation may not be the correct one.
3. Regardless of whether a suboptimal outcome is due to circumstances beyond physician's control or to a clearly inappropriate decision, a patient should be entitled to financial compensation.
4. Prosecution of individual physicians is justified only in rare cases of overt misbehavior.
5. Open discussion and careful analysis of every case of malpractice are essential steps for the prevention of it and the improvement of patient safety.

© Springer Nature Switzerland AG 2019
M. Zwitter, *Medical Ethics in Clinical Practice*,
https://doi.org/10.1007/978-3-030-00719-5_9

The *Nekrep Case* was a court trial that followed the death of 12-year-old Bor Nekrep, due to a misdiagnosis in Slovenia in March 2008. The treating pediatrician misdiagnosed an inborn defect of the urea cycle (a very rare genetic condition). After the boy had been brought by his parents to the Pediatric Clinic in Maribor, the doctor, among other things, measured the level of ammonia in his blood and on the basis of a significantly increased level erroneously diagnosed Reye syndrome. Because of the misdiagnosis and a delay in appropriate treatment, the boy developed cerebral edema and died 8 days later [1].

A family physician performs at least 5000—in some cases even more than 10,000—patient examinations per year. A surgeon performs over 100 major operations per year. An interventional radiologist working in a large medical center performs hundreds of catheterizations and coronary artery dilatations per year.

Is it reasonable to expect that a physician will never make a mistake in reaching a diagnosis, in the assessment of patient's health condition or in choosing the best mode of treatment? Is it realistic to expect that a surgeon or interventional radiologist will never fail despite their best intentions? We know the answer: it is not possible to completely avoid complications in medicine.

9.1   Mistake, Error, Neglect, Unfortunate Coincidence?

Some rephrase cases of medical malpractice as "complications," "error," "fault," "inaccurate judgment," "neglect," "unexpected consequences of treatment," "unfortunate coincidence," "poor communication among members of the medical team," or "grave professional neglect." In this chapter, we will deliberately leave such pigeonholing and euphemisms aside. From the patient's point of view, it, in fact, does not matter what the reasons were that led to a bad outcome of a medical procedure. Every individual patient has a right to expect optimal and safe medical treatment, and the medical system should meet such expectations. As we shall see, this is unfortunately not the case.

Rare are the countries with good systems of reporting and analyzing professional mistakes in healthcare. Sometimes physicians do not notice a mistake and fail to see that different measures might have prevented an unwanted course of a disease. Even more commonly, however, physicians remain silent about their own or a colleague's mistake since they wish to avoid criticism or even prosecution. Thus, the real burden of medical malpractice remains unknown.

9.1.1   Limited Resources, Compromises in Probability-Based Medicine, and the Right to Compensation

The three important implications of limited resources for medical care are an ethical and legal burden for physicians working under suboptimal conditions, poor efficiency of the existing financial resources, and suboptimal patient care. The first issue will be discussed here, leaving the other two for the next chapter.

In many countries with a healthcare system based on solidarity, we practice *probability-based medicine* since professional decisions are often subject to compromise. Physicians and other healthcare workers know very well what measures should be taken in any given situation; yet, optimal and professionally correct solutions are often not feasible due to financial limitations, limitations in personnel or due to long waiting periods. Compromises are most commonly made in diagnostics. Because there is no way we can perfectly verify a diagnosis, physicians must act probabilistically. Decision-making that follows the logic of sports betting is of course far from optimal.

In *probability-based medicine*, it is most often not possible to talk about individual responsibility. Better stated: it would be possible to address individual responsibility; however, one would have to go to a higher level—to a director of the hospital, to the national secretary for health, or to the prime minister. In any case, we cannot blame the physician as an individual. Consider the following concrete example:

> A trainee in carpentry had two of his right-hand fingers cut off during his traineeship. He immediately wrapped the fingers in a clean cloth and covered them with ice. He got to a hospital in 15 min. The only surgeon on duty had just started an extensive surgery of a patient who had been injured in a car accident. The surgeon was only able to start finger replantation procedure after three hours. After 2 weeks, it became clear that the reattachment of fingers failed which was certainly due to the overly long delay time between the accident and the surgery.

Is the surgeon to be blamed in this situation? No, of course not. Was patient management inappropriate and did the young man suffer damage? Yes, absolutely. Would he have the right to compensation? Well, now we find ourselves in trouble. Following the insurance practice in many countries, the patient is entitled to compensation only in cases when the physician is proven guilty. It would be fair that the insurance company (where every healthcare service provider is insured for cases of professional liability) paid the compensation. At the same time, the insurance company should inform the hospital management that they will raise the premium should similar cases reoccur. In such a way, the responsibility for poor professional service would shift towards the root of the professional mistake: the hospital management should reevaluate whether saving money on emergency services is justified.

9.2   Criminalization of Professional Malpractice

We have reached a crucial point: the criminalization of professional mistakes and the interdependence of the right to compensation and proven guilt are the biggest obstacles that prevent us from reporting, analyzing, and eliminating professional mistakes in patient management. It is clear that physicians will not report cases of incorrect diagnosis, inappropriate choice of treatment, or other types of malpractice if our self-reporting has financial consequences or could even bring us to court. The patient's right to receive compensation in cases in which no guilt is established is, therefore, crucial for physicians to be able to speak about mistakes, complications, or malpractice openly with patients as well as with their peers. We must focus on the event, not on the offender. Only once we accept this point of view will we be able to seriously analyze professional malpractice and prepare measures for the elimination of the underlying risk factors.

9.3   The Media, the Public, and Professional Mistakes

Many physicians are highly critical of the media that create sensationalistic reports of professional malpractice to increase their ratings or circulation. There is quite some truth to this criticism: supposed professional mistakes are often portrayed one-sidedly, disrespectfully or with contempt ("gods in white") along with the full disclosure of the physician's personal information. While credible media outlets should allow the other party—that is, the physician—to tell their part of the truth prior to publication so that both perspectives (the patient's and the physician's) are subsequently published, such a balanced presentation is rare. At the outset of the public discussion of a controversial case, physicians usually find themselves in a defensive position. It is similarly not fair that the patient or his lawyers can make public allegations whereas the physician's defense is limited by law due to the confidentiality of personal information. It was this reflection that led us to propose in the new version of the Slovenian Code of Medical Ethics the following: "Patients and their relatives determine the scope of confidentiality with respect to disclosing personal records. In cases of public accusation, the physician has the right to disclose those personal records that are crucial for an objective evaluation of the contentious act."

If we are sincere in placing the patient's interest first, we must collaborate responsibly with the media rather than turning our backs on them. To their credit, I should stress that without media reporting many controversial cases would never come to the light of day, numerous cases of professional malpractice would remain unidentified, and we would not be seeking ways to perform safer and professionally correct service. We have noted already how we physicians often conceal our professional mistakes: be it due to distress, guilty conscience, professional solidarity or due to the criminalization of professional malpractice. However, we are not alone in this game: have you ever seen an economist, a judge, or a politician publicly disclose their unlawful behavior at their own initiative?

It is unfortunate that other involved parties do not participate in the disclosure, analysis, and prevention of professional malpractice. For example, health insurance companies that fund the healthcare services should protect the interests of their insurance holders and carefully monitor the quality of the service provided.

9.4   Support for the Accused Physician

Finally, let us briefly discuss how to support the colleague who finds himself in distress due to allegations of professional malpractice. It is all too common that the accused physician is left to himself, alone against the public, the media, the lawyers, and even against his colleagues. Recall the story of Bor Nekrep, the unfortunate boy from Maribor whose portrait is shown as an illustration to this chapter. In this case, the hospital management kept silent for a long time, and renowned physicians publicly issued strong statements about the mistake. Even the Medical Chamber of Slovenia chose the wrong question regarding the appropriate medical procedure. Since their question was based on a subsequently established diagnosis, they hired an Austrian biochemist, a scientist, and an expert in a very rare metabolic disorder. Instead, they should have considered the situation when the boy was brought to the hospital and should have sought an answer about a professionally correct treatment of a 12-year-old febrile and confused boy. Let me be clear: I do not claim that the physician's management was professionally correct. I also do not defend her way of communicating with the child's parents, as presented to us in the media. However, I remain absolutely certain that everyone, especially the hospital management and her colleagues, should have offered her their support. Even if she, in fact, had committed a mistake, it is our duty to lend support to our colleague. Remember the words from the beginning of this chapter: there is none among us who have never made a professional mistake.

Reference

1. https://en.wikipedia.org/wiki/Nekrep_Case

Suggested Reading

Adina Kalet A, Chou CL. Remediation in medical education: a mid-course correction. Berlin: Springer; 2014. ISBN-13: 978-1493945061.
Bernstein M, Brown B. Doctors' duty to disclose error: a deontological or Kantian ethical analysis. Can J Neurol Sci. 2004;31:169–74.
Bono MJ, Hipskind JE. Medical malpractice. St. Petersburg, FL: StatPearls Publishing; 2018.

Durand MA, Moulton B, Cockle E, Mann M, Elwyn G. Can shared decision-making reduce medical malpractice litigation? A systematic review. BMC Health Serv Res. 2015;15:167. https://doi.org/10.1186/s12913-015-0823-2.

Kachalia A, Bates DW. Disclosing medical errors: the view from the USA. Surgeon. 2014;12:64–7. https://doi.org/10.1016/j.surge.2013.12.002.

Kassim PN. No-fault compensation for medical injuries: trends and challenges. Med Law. 2014;33:21–53.

Klaas PB, Berge KH, Klaas KM, Klaas JP, Larson AN. When patients are harmed, but are not wronged: ethics, law, and history. Mayo Clin Proc. 2014;89:1279–86. https://doi.org/10.1016/j.mayocp.2014.05.004.

Pellino IM, Pellino G. Consequences of defensive medicine, second victims, and clinical-judicial syndrome on surgeons' medical practice and on health service. Updat Surg. 2015;67:331–7. https://doi.org/10.1007/s13304-015-0338-8.

Perez B, Knych SA, Weaver SJ, Liberman A, Abel EM, Oetjen D, Wan TT. Understanding the barriers to physician error reporting and disclosure: a systemic approach to a systemic problem. J Patient Saf. 2014;10:45–51. https://doi.org/10.1097/PTS.0b013e31829e4b68.

Rodziewicz TL, Hipskind JE. Medical error prevention. St. Petersburg, FL: StatPearls Publishing; 2018.

Stamm JA, Korzick KA, Beech K, Wood KE. Medical malpractice: reform for today's patients and clinicians. Am J Med. 2016;129:20–5. https://doi.org/10.1016/j.amjmed.2015.08.026.

Surbone A. Onclogists' difficulties in facing and disclosing medical errors: suggestions for the clinic. Am Soc Clin Oncol Educ Book. 2012:e24–7. https://doi.org/10.14694/EdBook_AM.2012.32.e24.

Surbone A, Rowe M. Reflections of a colleague and patient on truth telling, medical errors, and compassion. Surg Oncol. 2010;19:185–7. https://doi.org/10.1016/j.suronc.2009.07.002.

Wienke A. Errors and pitfalls: briefing and accusation of medical malpractice – the second victim. GMS Curr Top Otorhinolaryngol Head Neck Surg. 2013;12:Doc10. https://doi.org/10.3205/cto000102.

Limited Resources, Priorities, and Corruption

10

Abstract

In this chapter, we discuss the situation in countries with a national healthcare system and do not cover countries with private healthcare insurance or those where services are paid out of pocket. Very few countries can meet all the costs of modern medicine for all their citizens. The three consequences of the gap between the available resources and the costs of modern medicine are limited access to costly medical interventions, waiting times for diagnostic and therapeutic interventions, and list of priorities. From the ethical standpoint, it is essential that all three of these issues are openly discussed and agreed within a wide circle including medical professionals and representatives of the lay community. In case of lack of democratic agreement and of clear rules, access to medical care may depend on the social and/or economic position of an individual. In a society based on solidarity, such a situation is contrary to the ethical principle of justice and may be considered to be a form of corruption.

© Springer Nature Switzerland AG 2019
M. Zwitter, *Medical Ethics in Clinical Practice*,
https://doi.org/10.1007/978-3-030-00719-5_10

Dr. Jerzy Einhorn was among the first to address the question of priorities in solidarity-based systems of health care (Photo: courtesy Dr. Stefan Einhorn)

At first glance, it seems unusual to include the organizational aspects of healthcare in a book on ethics. However, such an impression would be superficial; poor organization can place significant ethical burdens on all parties involved—patients, physicians, other healthcare workers, and society at large. If we carried out an international survey, negative points in almost all categories—autonomy, beneficence, non-maleficence, and justice—would be assigned to the organization of healthcare in many countries.

The central part of the discussion highlights ethical issues in countries whose legislation includes free healthcare for all their citizens. I am referring primarily to the East and Central European countries that have transitioned from single-party regimes with declared socialist ideologies to multiparty systems and a free-market economy but have at the same time maintained their old promises of a universal solidarity-based healthcare system. Ethically controversial organizational issues are less common in the West and North European countries with a longstanding tradition of democratic discussion about common needs. On the other extreme of the spectrum is the USA, where healthcare is predominantly based on corporate and profit-based principles. Such a system leads to very expensive healthcare which remains unaffordable to a significant proportion of the population.

No country in the world is rich enough to offer state-of-the-art and completely free universal healthcare. We cannot change the fact that there will never be enough money for healthcare. It is our ethical duty, however, to use the available means in the most rational way possible.

10.1   Consequences of Poorly Organized Healthcare

A thorough analysis of professionally and ethically questionable approaches in the organization, financing, and management of healthcare is, of course, beyond the scope of this book. Instead, we will focus on the difference between what is promised and what is realistically achievable. In all parts of the world, including former socialist countries, old Western democracies, and developing countries, this gap is widening. Since health insurance companies cover significantly fewer services than the citizens, in fact, need, waiting times arise and are becoming longer. Long waiting periods lead to negative consequences for the quality of patient care, for the rational use of available resources, and may lead to corruption.

In cases of long waiting times for up-to-date diagnostics, a definitive diagnosis cannot be established. Instead of a firm diagnosis, physicians opt for the most probable one. This leads to the practice of *probability-based medicine*. As we said in the previous chapter on malpractice, decision-making that follows the logic of sports betting is far from optimal. If not treated immediately, many diseases deteriorate, leading to inferior service to patients and to additional costs for treatment of more advanced disease. Waiting periods in diagnostics and treatment also prolong sick leave.

Insufficient public funding and the resulting waiting times open the doors for out-of-the-pocket immediate services for procedures that should have been available within a reasonable time to anyone. Self-funding in private healthcare is in serious conflict with solidarity and the ethical principle of justice. Waiting periods come with an additional problem that the wealthy patients and those with influential acquaintances and connections will find a way to faster diagnostics and treatment. Circumvention of waiting periods is frequently the most significant form of corruption in healthcare.

Members of disadvantaged communities are those who are most significantly affected by the poor organization in healthcare. The socially disadvantaged and those with no social influence cannot circumvent waiting periods and have substantial difficulties in affording private healthcare service.

In the background of discussions on waiting periods and on the accessibility of healthcare is the question of the rational use of healthcare resources. As already said, a delay in medical intervention means that a disease is more advanced and hence the treatment is more expensive. Additional unnecessary costs arise from the fact that patients with minor complaints can wait and will ultimately get the service for free. This leads to unnecessary specialist examinations even for minor issues, hospitalization due to social reasons, and originator instead of generic drugs.[1] In the non-rational use of healthcare means, physicians play their part as well. Too often, they are not critical in the evaluation of reports from sponsored clinical trials and insist on the introduction of new costly drugs which may be only of marginal benefit in comparison to old and well-established therapies.

We can now reaffirm our opening remark in this chapter: poor organization in healthcare places significant ethical burdens, from the perspective of beneficence,

---

[1] Originator (also original) drugs are those that have been granted marketing authorization on the basis of studies done by the manufacturer that prove the drug's safety, quality, and efficacy. Once the originator drug loses patent protection, another manufacturer can make and market a generic drug with the same active substance. Generic drugs are typically not tested in additional clinical studies and can be significantly cheaper.

non-maleficence, and justice, on all those involved—patients, healthcare workers, and society at large.

From a principled perspective it is clear: in solidarity-based healthcare systems, citizens should be promised only those healthcare services that can be realistically performed under the state budget. It is, therefore, necessary to determine and agree on priorities, a topic of later sections in this chapter.

10.2   Probability-Based Medicine

In many countries with solidarity-based healthcare systems, we practice *probability-based medicine*. Due to waiting periods and because not all contemporary diagnostic tools are available, the physician must frequently speculate about the most likely diagnosis. True, the physician can tick "urgent"[2] on the referral document for cases requiring examination on the same day; however, the physician could have done the same for all the patients. Technically, it is clear that the level of urgency cannot be reliably determined until we have established a firm diagnosis. Any form of chest pain can be due to heart infarction; lower back pain or problems with digestion can be the first signs of cancer. The same holds for treatment; the physician does not always have all the means available to perform a safe and the most successful treatment. Consider the example of managing labor pain: technically, the best solution is to use epidural analgesia. Nevertheless, labor pain is still frequently managed with Remifentanil, which is less efficient and can cause respiratory depression in newborns.[3] Obstetricians are well aware of this issue—but what should they do when there is no anesthesiologist at hand who could perform epidural analgesia?

Probability-based medicine is the source of at least three significant deficiencies in the healthcare system: it curtails the levels of professionalism, imposes ethical and legal responsibilities on physicians, and leads to increased costs. The significance of consequences that the lack of the latest diagnostics and treatment has for effective medical treatment cannot be overstated. Similarly, it is evident that we impose a significant ethical burden and also a legal responsibility on the physician who cannot follow his expertise in patient management. However, we are quick to forget that attempts to reduce the costs of modern medical treatment can easily backfire. The initial, properly planned, and correctly performed treatment is often significantly cheaper than the inappropriately chosen treatment with the subsequent management of associated complications. Imagine the costs that the family and the health insurance company have to bear because a child in the family is suffering

---

[2] In Slovenia, the physician can indicate one of the four levels of urgency when referring a patient to a specialist: urgent, very fast, fast, or regular.

[3] In the drug information leaflet, the manufacturer writes: "We currently do not have sufficient data based on which we could recommend the use of Remifentanil during labor or Caesarean section. Remifentanil is transported through the placental membrane; analogues of Remifentanil can cause respiratory depression in newborns." English anesthesiologist Prof. Felicity Reynolds Spencer, a leading expert, wrote that Remifentanil is "worse than useless" for managing labor pain and in many respects harmful to the newborn (personal communication, January 2, 2016).

from cerebral palsy—a situation that could have been avoided if delivery at childbirth had been performed professionally. Or take the example from my own area of expertise: in about a third of lung cancer patients with an apparently localized disease, the PET-CT examination shows that the cancer has spread widely. In such situations, aggressive surgical treatment or radiation therapy are not appropriate. However, because we are frequently forced to make decisions without performing the PET-CT examination, some patients are subjected to inappropriately aggressive treatment which also represents huge and unnecessary costs. Indeed, in discussions on probability-based medicine, on waiting periods, and on the availability of diagnostics and treatment, a bit of mathematics and utilitarian ethics is needed.

10.3   Priorities

The years are passing by quickly; one health reform will be followed by another. Sooner or later, we will realize that we must give up certain rights in healthcare and promise people only what we can pay for. However, whatever is promised should be available without waiting periods, out-of-pocket payment, and corruption. When this time comes, we will be faced with a difficult task of defining our priorities.

At present, solidarity-based healthcare systems are still prevalent in Europe, where we believe that people must not be discriminated on social grounds when it comes to disease. Such understanding of justice in health and disease is evidently beneficial to the disadvantaged part of the population that on average gains significantly more from the healthcare system, compared to its financial contribution. An additional advantage of solidarity-based systems is that the more influential classes of society support a decent level of public healthcare; otherwise, they would be going against their own interests of receiving proper treatment in case of a disease. The downside of this type of systems is the recognition that no country on earth can offer completely free, state-of-the-art healthcare. Even the rich countries of the Persian Gulf guarantee free healthcare only to its citizens but not to the numerous foreign immigrants on temporary work visas. In solidarity-based healthcare systems, we are, therefore, inevitably faced with the problem of priorities: we must determine what services are included in healthcare and are provided to all citizens and what cannot be provided.

Scandinavian countries and Great Britain are examples of countries whose politicians had sufficient courage and honesty and stated clearly what the limits of the healthcare system are. The citizens are not promised "castles in the sky," but only as much as the state can afford. They have a system of priorities—and therefore also a list of healthcare services that their public healthcare systems cannot offer. The cost efficiency of specific healthcare measures is not the only criterion; however, it is a meaningful one when such national programs are being developed. Ethical aspects of healthcare measures are also significant.

Defining national priorities in healthcare is far from trivial. There are at least two reasons why we must not entrust defining of priorities entirely to the physicians and the healthcare sector alone. First, it would provoke serious disputes among physicians; every physician will care primarily for his own speciality. Who is to adjudicate if dermatologists, orthopedist, and oncologists start competing for inclusion on the priority list? The only way out of such disputes would be a rigorous application of utilitarian ethics and calculation of costs per year of life gained including the additional factor of quality-adjusted life years. However, this approach in itself is difficult since we do not have accurate data about the efficiency of specific modes of treatment and because there can be huge disagreements in estimating quality-adjusted life years. Second, we cannot entrust the selection of priorities exclusively to physicians since the consideration necessarily includes other values not based on medicine. We physicians should be aware that we might know how to manage diseases; however, everyone has the right to define the values upon which our decisions are based.

Jerzy Einhorn, a Polish Jew by birth, was one of the most respected Swedish oncologists and a member of the Swedish parliament in his later years. In 1993, the Swedish healthcare authorities realized that, in spite of their high living standard, they could no longer provide to their people all services of contemporary medicine. Dr. Einhorn then led a parliamentary committee, travelled even to the most remote parts of the country and conducted four extensive surveys. When deciding on priorities in healthcare, they followed the three core principles: human dignity, human needs and solidarity, and cost efficiency. On the basis of numerous discussions involving several thousands of people from all strata of society, the Swedish Priority Commission proposed the following priority groups in clinical activity (Table 10.1).

In many Central and East European countries, we adhere to unlimited rights which in practice, however, remain unavailable to the majority of people. We currently do not yet dare to raise the issue of setting priorities in healthcare and limiting the list of healthcare services. At the same time, we should agree that minor, less important healthcare services should be paid directly by the citizens themselves or through voluntary, complementary health insurance.

Agreement about healthcare priorities requires much political courage. The biggest barrier towards reaching a social consensus is the political decision. Politicians

Table 10.1 Swedish Priority Commission: priority groups in clinical activity [1]

Priority group	Content of care
IA	Care of life-threatening acute diseases and diseases which, if left untreated, will lead to permanent disability or premature death
IB	Care of severe chronic diseases
	Palliative terminal care
	Care of people with reduced autonomy
II	Individualized prevention in contacts with medical services
	Habilitation/rehabilitation as defined in the Health and Medical Services Act
III	Care of less severe acute and chronic diseases
IV	Borderline cases
V	Care for reasons other than disease or injury

should honestly tell the citizens that public healthcare insurance does not cover the costs of all medical procedures. Understandably, many will have little enthusiasm for such a message. Plans for a broad social and political consensus on the necessity of healthcare reform are diluted when concrete solutions are proposed. Healthcare reforms are listed on agendas of almost all governments in Central and East European countries; however, I cannot escape the impression that these are mostly cosmetic changes with which the politicians buy themselves time to the next election. The politicians are not addressing the crucial question—redefinition of the promised rights. Therefore, we cannot expect improvement in access to healthcare services and alleviation of all professional, ethical, and also financial problems that arise due to the discrepancy between what is promised and what is realistically achievable.

10.4 Corruption

Let us open the discussion on corruption with the issue that we have already discussed in some depth in this chapter: waiting periods. Upon receiving information about the estimated waiting time for diagnostics or treatment, the patient has three options. First, he can practice patience and wait. Second, he pays for the healthcare service himself as a self-funded patient. Third, he can seek ways of bypassing the waiting list. In the latter example, the patient, therefore, steps ahead of all other patients on the waiting list. From an ethical standpoint, this is undoubtedly a form of corruption, regardless of whether such an arrangement is due to personal acquaintances or bribery.

The chances are slim that stricter forms of control could eliminate the circumvention of waiting lists. Medicine is not mathematics, and it is challenging to precisely triage the patients according to the degree of urgency in each medical intervention. Stricter control over waiting periods also leads to additional ethical issues. Consider the example of treating a medical colleague: without exception, all ethical codes from the Hippocratic Oath onwards assign the special duty of attention and solidarity when offering medical help to other physicians. In contemporary codes of medical ethics, the scope of solidarity is often broadened to include other healthcare workers and medical students. Another example of an ethically justified criterion for preferential treatment is the patient's age. Let me illustrate this with an anecdote from my professional practice. When a colleague of mine called and asked me for the admission of a 28-year-old lung cancer patient, I accepted her the very next day; an elderly patient with the same diagnosis got an appointment a week later. The decision on preferential treatment of a diseased physician or consideration of the patient's age as the criterion for preferential treatment nicely highlights the dilemma discussed in the introductory chapter on the relationship between law and ethics: although such decisions are legally debatable, they are nonetheless ethically correct. We can, therefore, conclude: the ethical burden of waiting periods can be alleviated only by ensuring that the services that the patient needs are available within a reasonable time.

The second field of possible corruption in healthcare is the involvement of physicians in the provision of medical materials and equipment. Physicians, of course, must collaborate in these decisions. However, it is essential that the criteria for selection are impartial and defined in advance. Any form of direct or even indirect profits by physicians or others involved in the procurement process is ethically and legally absolutely unacceptable.

The third area of corruption in healthcare opens up whenever a physician plays a double role—as a physician who has the duty to treat the patient and as the owner, co-owner, or a shareholder in a healthcare institution. In such cases, physician's decisions are not only influenced by considering patient's interests but also by commercial interests. The experience from the USA shows that this is particularly evident in choosing between more or less expensive drugs. Manufacturers will offer a significant discount on the purchase price for the newer, expensive drugs whereby the hospital will charge the insurance company the full price of the drug. Such price differences, and hence profit margins, are however not available for older medications, especially for drugs whose patent protection has already expired. Older and cheaper drugs with which we have had a long-lasting good experience are falling out of the market because they do not bring profits to the manufacturers nor to the healthcare institutions. It, therefore, does not come as a surprise to learn that the USA has the highest healthcare costs per capita per year globally and at the same time a large number of its citizens do not have free access to basic healthcare.

Reference

1. Calltorp J. Priority setting in health policy in Sweden and a comparison with Norway. Health Policy. 1999;50:1–22.

Suggested Reading

Ciapponi A, Lewin S, Herrera CA, Opiyo N, Pantoja T, Paulsen E, et al. Delivery arrangements for health systems in low-income countries: an overview of systematic reviews. Cochrane Database Syst Rev. 2017;9:CD011083. https://doi.org/10.1002/14651858.CD011083.pub2.
Cromwell I, Peacock SJ, Mitton C. 'Real-world' health care priority setting using explicit decision criteria: a systematic review of the literature. BMC Health Serv Res. 2015;15:164. https://doi.org/10.1186/s12913-015-0814-3.
Fischer T, Langanke M. Individualized medicine: ethical, economical and historical perspectives. Berlin: Springer; 2015. ISBN-13: 978-3319346984.
Gaitonde R, Oxman AD, Okebukola PO, Rada G. Interventions to reduce corruption in the health sector. Cochrane Database Syst Rev. 2016;8:CD008856. https://doi.org/10.1002/14651858.CD008856.pub2.
Goldberg DS. Public health ethics and the social determinants of health, Springer Briefs in Public Health. Berlin: Springer; 2017. ISBN: 978-3-319-51347-8.
Gu Y, Lancsar E, Ghijben P, Butler JR, Donaldson C. Attributes and weights in health care priority setting: a systematic review of what counts and to what extent. Soc Sci Med. 2015;146:41–52. https://doi.org/10.1016/j.socscimed.2015.10.005.

Jamison DT, Alwan A, Mock CN, Nugent R, Watkins D, Adeyi O, et al. Universal health coverage and intersectoral action for health: key messages from disease control priorities, 3rd edition. Lancet. 2018;391:1108–20. https://doi.org/10.1016/S0140-6736(17)32906-9.

Kagawa-Singer M, Dadia AV, Yu MC, Surbone A. Cancer, culture, and health disparities: time to chart a new course? CA Cancer J Clin. 2010;60:12–39. https://doi.org/10.3322/caac.20051.

Kakuk P, editor. Bioethics and biopolitics: theories, applications and connections, Advancing Global Bioethics. Berlin: Springer; 2017. ISBN: 978-3-319-66249-7.

Kantarjian H, Rajkumar SV. Why are cancer drugs so expensive in the United States, and what are the solutions? Mayo Clin Proc. 2015;90:500–4. https://doi.org/10.1016/j.mayocp.2015.01.014.

Knaul F, Horton S, Yerramilli P, Gelband H, Atun R. Financing cancer care in low-resource settings. In: Gelband H, Jha P, Sankaranarayanan R, Horton S, editors. Cancer: disease control priorities, vol. Vol. 3. 3rd ed. Washington, DC: The International Bank for Reconstruction and Development/The World Bank; 2015.

Moses H 3rd, Matheson DH, Dorsey ER, George BP, Sadoff D, Yoshimura S. The anatomy of health care in the United States. JAMA. 2013;310:1947–63. https://doi.org/10.1001/jama.2013.281425.

Nass SJ, Madhavan G, Augustine NR, editors. Making medicines affordable: a national imperative. Washington, DC: National Academies Press; 2017.

Patel V, Parikh R, Nandraj S, Balasubramaniam P, Narayan K, Paul VK, Kumar AK, Chatterjee M, Reddy KS. Assuring health coverage for all in India. Lancet. 2015;386:2422–35. https://doi.org/10.1016/S0140-6736(15)00955-1.

Prainsack B, Buyx A. Solidarity in biomedicine and beyond. Cambridge: Cambridge University Press; 2017. ISBN-13: 978-1107074248.

ten Have H. Vulnerability: challenging bioethics. Abingdon: Routledge; 2016. ISBN-13: 978-1138652675.

Image of a Physician

11

Abstract

The history of medicine and of humanity contains a long list of exceptional physicians who devoted their lives to alleviating human suffering, often in dismal conditions, at risk to their own health and life. However, these outstanding personalities should not be taken as a model towards whom every physician should strive. For most physicians, their profession is much more than just a job; yet, they do not wish to be deprived of their private lives. Indeed, family with its pleasures and obligations, culture, sports, or any other interests are vital to counterbalance the stress of medical profession and prevent or mitigate burnout. In concluding this chapter, comments are offered on some problematic topics: physicians who share their work between public institutions and a private practice; physicians who perform scientifically unproven treatment; and medical strikes as a method to press for economic and other interests of the medical profession.

Dr. Dorothy Reed (1874–1964; upper left panel) was among first women to complete studies of medicine. She is remembered for two outstanding achievements: a pivotal paper on the biology of Hodgkin's disease, written at the age of 28 when she was a resident in pathology; and a most influential study on infant and maternal mortality rate which led to a drastic change in the practice of perinatal care in USA ([1]; Photo courtesy Dr. John T. Mendenhall)

Dr. Magomed Gadžijev (1914–1972; upper right panel), from Dagestan, served as a surgeon in the Red Army. He was severely wounded in a battle in Ukraine and captured by German forces. After escaping from a German prison, he joined partisans in Slovenia and was in charge of a partisan hospital in Zalesje

Dr. Franja Bidovec (1913–1985; lower left panel) was in charge of the partisan hospital which operated during the Second World War in a remote valley near Cerkno, Slovenia. Almost 600 severely wounded persons were treated in what is now known as Franja Partisan Hospital and is on the UNESCO List of World Cultural Heritage

Dr. Catherine Hamlin, born 1924 (lower right panel) is an Australian gynecologist and obstetrician who spent most of her career in Ethiopia. Catherine and her husband Dr. Reginald Hamlin founded the Addis Ababa Fistula Hospital, the world's only medical center dedicated exclusively to providing free obstetric fistula repair surgery to poor women suffering from childbirth injuries. In 2014, Ethiopia nominated her for the Nobel Peace Prize. (Photo: Joni Kabana)

Recently, I was talking to cum *laude* graduates. Quite a few of them planned to study medicine, and all were talking about their future profession with incredible enthusiasm. The young are aware that being a physician is not "just another job" allowing you to live a more or less comfortable life. Although we will briefly touch upon the disappointment experienced by physicians, it must be emphasized that the exceptional dedication in pursuing their profession that is seen in medicine is not present in any other profession. As individuals and as members of our professional associations, physicians are constantly protecting the interests of our patients. What is perceived as the norm by physicians is a rare exception in other professions. You will rarely meet an economist or a lawyer who will show similar levels of dedication in fighting for the rights of their workers or for fair treatment of their business partners.

11.1 The Physician in Exceptional Circumstances

Albert Schweitzer,[1] Franja Bidovec,[2] Magomed Gadžijev,[3] Catherine Hamlin,[4] Lev Milčinski[5]—these are just some of the names of exceptional physicians. Alongside theirs, we could have listed the names of thousands of physicians who dedicated their lives to their profession. They dismissed the risks to their health, they did not pay attention to time and weather, their annual vacation, paycheck, or even their families while they patiently—and frequently with heaps of goodwill—carried the burden of the medical profession. Experience of serving severely injured patients or those with contagious diseases, sleepless nights, and kilometers of snow and mud for the spark of gratitude in the eyes of the patient. Each of them deserves to have a book written, or film made about them.

A visit to the Franja Partisan Hospital[6] offers an unrivalled experience. Physicians, especially those working in well-equipped university clinical centers, find it hard to imagine the amount of enthusiasm, inventiveness, and courage that was required to

---

[1] Dr. Albert Schweitzer: French-German theologian and physician, the founder of the hospital in Gabon, Nobel Peace Prize laureate.

[2] Dr. Franja Bidovec: physician who worked at the Franja Partisan Hospital which was set up in a hidden gorge near Cerkno, Slovenia. See also the seminar "Franja Partisan Hospital."

[3] Dr. Magomed Gadžijev: surgeon from Dagestan, the founder of the Zalesje Partisan Hospital.

[4] Dr. Catherine Hamlin: gynecologist from Australia who founded a hospital for treatment of birth traumas. See also the student seminar paper on "Catherine Hamlin."

[5] Dr. Lev Milčinski: Slovenian psychiatrist and university lecturer.

[6] Franja Partisan Hospital: a secret World War II hospital near Cerkno in western Slovenia. It was run by the Slovene Partisans from December 1943 until the end of the war as part of a broadly organized resistance movement against the Fascist and Nazi occupying forces. The wounded being treated there were soldiers from both the Allied Powers and the Axis Powers. Although the occupying Wehrmacht forces launched several attempts to find the hospital, it was never discovered. Today it exists as a museum. It has been protected as a cultural monument of European significance.

perform such work. The military hospital, set deep in a mountain gorge, even had an X-ray machine, and there was no waiting list for it!

Extraordinary life circumstances can elicit surprising capabilities in human beings, for better or for worse. In war, poverty, hunger, or disease, some become capable of killing while others develop the incredible will to help people in existential distress. While these achievements are deservedly praised, it should be added that these individuals were also lucky that their fate led them to the circumstances in which their heroic actions could be recognized. Or to phrase this differently: a contemporary physician is no less ethical even though he is "merely" a family physician working in a remote rural area.

This brings us to our next question:

11.2   Who Is a Good Physician?

I recently listened to Vesna Milek's[7] interview with the Slovenian Cardinal Franc Rode.[8] When asked about the avowed celibacy of Catholic priests, the Cardinal offered examples of Mother Teresa and the missionary Pedro Opeka.[9] "Can you imagine Mother Teresa as a married woman or Pedro Opeka as a married man? Would they still be able to follow their calling and perform their mission?" The answer contains a logical flaw: indeed, I find it difficult imagining Mother Teresa and Pedro Opeka as married; yet, I have no trouble imagining a good priest who is also a good father. Celibacy is not problematic as long as it is not imposed. In other words: we must not lay down rules according to the characteristics of exceptional individuals.

A great majority of physicians are devoted to their profession, yet, they do not wish to dismiss family life, friends, culture, and sports. Our discussion of exceptional physicians from the beginning of the chapter should not be seen as an appeal that all physicians should follow the examples of the greats.

Since almost all the chapters of this book provide ethical guidelines for physicians, it makes little sense to repeat ourselves here. Let us state briefly that a good physician is not a superman, but he is still more than just an employee. The physician understands his vocation as a commitment to help fellow human beings in distress. A good physician must be a qualified expert who is at the same time kind and compassionate, and who knows how to adapt the communication to each patient and his family. Regardless of all the flaws in the healthcare organization in which he works, he will not forget his mission. When an additional patient is brought to his already overcrowded office, a compassionate physician will not close the door.

---

[7] Vesna Milek: Slovenian writer and journalist.

[8] Dr. Franc Rode: Slovenian archbishop and cardinal.

[9] Pedro Opeka: Slovenian missionary, born in Argentina. He was nominated for Nobel Peace Prize for his work on improving living conditions in Madagascar.

11.3 Who Is Draining the Idealism from Our Profession?

In 6 years of study, medical students are taught to perform the best healthcare measures: the best ways to prevent disease, the best diagnostics, and the best treatment. These students do not hear about physician's time limitations, about substandard hospital facilities or financial restrictions. Students do not receive lectures on poor healthcare organization, which was discussed in the previous chapter.

When a recent graduate puts on the doctor's uniform, he suddenly starts seeing the profession of the physician in a different light. He gradually recognizes that he is not working in an isolated world, where his only task is to make decisions that benefit the patient entrusted to his care. Instead, he is working in a real world of interconnected influences, interests, and restrictions. The decisions about preventive measures, diagnostics, and treatment are therefore not based solely on medical expertise but are very often a compromise between the ideal and the realistically doable. The idealism from the student years wanes quickly and is displaced by uncertainty, fear, and concerns over one's future.

A young physician at the beginning of his professional path faces three problems. The first is the discrepancy between the promised best healthcare and what the healthcare system can realistically offer to people. The second problem is the consequence of the first: optimal medical treatment is being increasingly displaced by *probability-based medicine*. The third problem is the idealization of the physician as an infallible, almost perfect expert. The common point in all these problems is that in cases of patient's or relatives' dissatisfaction, the physician is left by himself, publicly exposed on the pillory of the media, with widespread accusations, and even prosecution in courts.

We have created a healthcare system with an ideal façade that conceals its numerous weaknesses. It is time to state clearly that it is not fair that the individual physician pays the cost of all financial restrictions and organizational flaws in healthcare. Those in charge have to accept their share of responsibility if we want physicians to regain their self-confidence, professional competence, and honest communication and collaboration with patients. The consequence of such fake ignorance by the healthcare management is that the entire burden of unrealized and unrealizable healthcare rights is placed on individual physicians. Similarly, individual physicians are the only ones who carry the whole burden in cases of public accusation of medical malpractice. Their disappointment is even deeper if their colleagues do not stand by their side in such cases.

11.4 Burnout and the Anchor Outside Medicine

Burnout is a serious, but often overlooked companion of the medical profession. Physical, mental, and emotional burdens accumulate over the years and affect many among us. A successful professional path of a physician can suddenly decline. Professional burnout also affects the family relationships. Many physicians do not recognize their own problems. Ignorance of personal vulnerability may lead to alcohol and other drug abuse. Suicide is also not rare.

As for most phenomena in nature, in society, and in most diseases, we must not look for a single causal factor of burnout. Work overload is but one of the causes. At least of equal importance is dissatisfaction with the work environment. The loss of idealism described in previous paragraphs contributes significantly to the feeling of powerlessness and work dissatisfaction; in such situations, burnout is just a small step away. Burnout prevention should therefore not be sought only through work regulation and workload restrictions, but also in ensuring a stimulating work environment. Burnout is not only a problem of the individual, but it also points to poor relations in the team.

An essential antidote to burnout is the world outside medicine. Family, friends, art, sport, love for nature, travel—every physician must have a world outside his profession. Active forms of pastime are the anchor to which the physician can attach himself when professional problems become unbearable.

11.5   Conscientious Objection and the Respect of Personal Values

We most commonly think of conscientious objection in relation to abortion, especially abortion without a medical indication. Some physicians extend conscientious objection to other medical procedures related to conception and pregnancy: contraception and procedures for physician-assisted insemination. In countries with legalized euthanasia, some physicians declare conscientious objection against performing euthanasia.

Any form of legal or ethical constraints on the personal values of an individual, be it physician, patient, or a healthy person, would be unacceptable. We do not hide our personal beliefs. The physician should be allowed to present his attitude towards particular medical interventions in an open conversation with co-workers, patients, or the public. Constitutional and legal provisions on the right to freedom of choice in family planning should not be interpreted as an obstacle to medical counselling. When approached by a pregnant patient who wants to carry out an abortion, the physician has the right and duty to express his view and may present alternative options such as an application for financial support or giving the baby up for adoption. The physician is therefore not an automaton blindly following the law or wishes of the person in front of him. At the same time, however, he also respects differences in opinion, and he should not impose his beliefs.

Even though it is not acceptable to deny physicians the right to conscientious objection, we must be aware that conscientious objection in some cases can lead to disagreements or conflicts in allocating the performance of certain tasks. It would be wrong to conclude that some physicians "have a conscience" and others "do not have" it. A few years ago, a specialist in gynecology and obstetrics and working in primary care declared conscientious objection and refused to prescribe contraception or give advice regarding physician-assisted insemination. While we have to respect her personal belief, she should also have recognized that her beliefs were in clear conflict with those of a large majority of the women attending her out-patient service. She should have understood that even among religious women, the large majority use contraception; and that artificial insemination brings happy family life to many who are not able to conceive by natural means. In my opinion, this physician should have chosen a different specialization, or she should have worked with elderly gynecology patients.

11.6 Digressions

Virtually all medical codes include a variant of the sentence: "The physician shall never use the patient for his own interests." With this sentence in mind, I will evaluate three phenomena: conflict of interest for physicians who offer their service both in public institutions and as private physicians, conflicts between physicians, and medical strikes.

The coexistence of public and private healthcare service providers is a reality which we cannot and also do not want to give up. Private healthcare service providers are introducing new services to healthcare, invest their own money in infrastructure and equipment, and are often exemplary in terms of well-managed service and the relationships with patients. It becomes problematic, however, when the same physician is working in a public institution and is additionally running his private office. In such cases, it is hard to avoid conflicts of interest. Will the physician really work with the same enthusiasm for the public institution if he knows that he has work to do in his private office in the afternoon? Can we prevent the shifting of patients from public institutions to private offices and back? Can we ensure that private and public institutions carry an equal burden concerning education, sick leave, and annual leave?

In a recent interview, Nobel laureate Professor Harald zur Hausen pointed to another phenomenon, also related to sharing the work between a public institution and private practice. Due to their additional well-paid work in private practice, physicians from university clinics do not engage in clinical and laboratory research. Because of this situation, the mission of university clinics as centers of independent research is severely endangered.

Due to negative aspects and potential conflicts of interest, I am not in favor of a dual role of physicians—employment in a public hospital and engagement in private practice. However, those employed in public institutions should receive additional remuneration for performing work that goes beyond the scope of regular work.

Conflicts among physicians about professional or organizational issues are common and should be regarded as normal. As we stressed before, medical decisions often depend on our values, and it is understandable that similar situations may be dealt with in a different manner. When a conflict among physicians reaches beyond the levels of interpersonal conversation, conflict resolution becomes the responsibility of the higher level on the organizational scheme. The basic guidelines are tolerance within the confines of what is deemed professionally and ethically accept-able, and the absolute protection of patients' interests.

Finally, a word on medical strikes. I know that many physicians experience excessive workload and that, in particular, the work of young physicians is under-paid. Physicians should be allowed to raise these issues in public. In every country, the voice of physicians is heard. The basic rule for any medical protest should amount to keeping the patients safe and preventing public opinion from turning against us. It is very problematic, however, if our demands are voiced in the form of a strike. Some years ago, German physicians in public hospitals organized an inter-esting form of protest: they continued with patient care but did not fill out administrative forms. As a consequence, healthcare management could not charge the costs to insurance companies. Such a form of protest is ethically preferable to a strike, which always affects patients.

Reference

1. Zwitter M, Cohen JR, Barrett A, Robinton ED. Dorothy Reed and Hodgkin's disease: a reflection after a century. Int J Radiat Oncol Biol Phys. 2002;53:366–75.

Suggested Reading

Brazier MR, Gillon R, Harris J. Helping doctors become better doctors: Mary Lobjoit—an unsung heroine of medical ethics in the UK. J Med Ethics. 2012;38:383–5. https://doi.org/10.1136/medethics-2012-100578.
Galton DJ. Is medicine still a profession? QJM. 2015:1–6. https://doi.org/10.1093/qjmed/hcv162.
Rothenberger DA. Physician burnout and well-being: a systematic review and framework for action. Dis Colon Rectum. 2017;60:567–76. https://doi.org/10.1097/DCR.0000000000000844.
Shanawani H. The challenges of conscientious objection in health care. J Relig Health. 2016;55:384–93. https://doi.org/10.1007/s10943-016-0200-4.
Surbone A. Professionalism in global, personalized cancer care: restoring authenticity and integrity. Am Soc Clin Oncol Educ Book. 2013:152–6. https://doi.org/10.1200/EdBook_AM.2013.33.152.
Thompson SL, Salmon JW. Physician strikes. Chest. 2014;146:1369–74. https://doi.org/10.1378/chest.13-2024.

Preventive Medicine

12

Abstract

Vaccinations, medical counselling for children and adults with the promotion of healthy life habits, and programs for the early detection of chronic diseases, such as some forms of cancer, cardiovascular diseases, and diabetes, are saving more lives than any branch of curative medicine. Preventive medicine is the main factor in prolonging life expectancy and is cost-effective. In ethical consideration, however, we should not forget that the activities of preventive medicine are directed towards healthy people. Any decision for the unsolicited intrusion of medicine into the lives of healthy persons should be made only after very thoughtful consideration, including public debates with representatives of non-medical professions and promotion of preventive programs in lay media. Apart from actions directed towards individuals, preventive medicine also has a vital role on the population level. Opposing all attempts to silence their voice, physicians should retain their professional integrity and should analyze and openly discuss issues that significantly impact human health, such as poverty, environmental pollution, pollution at work, and all kinds of addiction.

© Springer Nature Switzerland AG 2019
M. Zwitter, *Medical Ethics in Clinical Practice*,
https://doi.org/10.1007/978-3-030-00719-5_12

Environmental pollution is a major risk factor for a wide spectrum of chronic diseases. Pollution and its companion, global climate change, are becoming a major threat to humanity

At a medical congress in St. Gallen, Switzerland, I heard the following anecdote:

> "Help!" Many ran down to the river and started rescuing the people who were being carried away by the fast-flowing river. Suddenly, one of the rescuers left. The others asked him, baffled, whether he would no longer help in rescuing. "No," he replied, "I will go and see who is throwing our people in the water." He followed the river upstream to its source and killed the dragon.

Better living conditions, hygiene, vaccination programs against contagious diseases, and antibiotics in curative medicine—these advances have contributed more to the prolongation of life expectancy than anything in the rest of medicine. The role of preventive medicine for public health is undisputed. However, with regard to the ethics of preventive medicine, it is nevertheless reasonable to have certain reservations. The physician must be aware that in such situations he is not dealing with a patient seeking help, but with a healthy person who has not asked for an opinion or for someone to interfere with his or her lifestyle.

12.1   Vaccination

We will discuss vaccination in greater depth in the chapter on pediatrics, especially in relation to parents who oppose the vaccination of children. Legislation enforces vaccination against those contagious diseases when direct contact with a sick person leads to acute disease and when vaccination offers reliable protection. In this chapter, we will consider the cases of diseases for which vaccination is not obligatory, that is when individuals voluntarily opt for vaccination. Seasonal flu is an example of a disease for which vaccination does not offer reliable protection. Vaccination against the papillomavirus is also not mandatory. Infection is most commonly transmitted during sexual intercourse and, after a long-lasting latent phase, cancer may develop in a relatively limited number of infected patients. Finally, there are also diseases that are not transmitted from human to human, such as tick-borne encephalitis, for which the decision for vaccination is similarly left to each individual.

To make the discussion concrete, we will consider as examples vaccination against tick-borne encephalitis and human papillomaviruses.

In Slovenia, every year 100–300 people contract tick-borne encephalitis of whom a significant number are left with permanent disabilities, and one or two may die. Tick-borne encephalitis is, therefore, an important healthcare issue with significant treatment costs. While Slovenia is among the countries with the highest proportion of infected ticks and with a high prevalence of the disease, the vaccination rate is below 10%. The reason for low immunization rates is clear: more than a 100 euros for three shots of the vaccine, 500 euros for a family of five. Free vaccination programs have been prepared, yet they are constantly shifted from one filing cabinet to another, passed from one minister to another. Is there truly no one who could bring this unethical apathy to an end? Someone who would press the vaccine providers to lower the price and who would promote free vaccination? We do not need to look very far for examples of good practice: in neighboring Austria, free vaccination and an accompanying information campaign led to an 80% immunization rate and a significant reduction in the numbers of patients with encephalitis.

Free vaccination programs are examples of cost-efficient preventive measures.

Human papillomavirus is the main cause of cervical cancer and to a lesser extent also of laryngeal and esophageal cancers. Results of large studies confirm that in girls, vaccination against the virus provides effective protection against infection and significantly reduces the risk of later cancer development. In Slovenia, vaccination against human papillomavirus is free of charge, but voluntary, which is why many parents avoid it. If they trusted the scientific evidence rather than misleading information by the opponents of vaccination, we could have prevented approximately ten deaths per year.

12.2 Healthy Lifestyle

An essential area of preventive medicine is the promotion of healthy lifestyles. Unhealthy diet, obesity, smoking, alcohol, illegal drugs, stress, poverty, pollution in work and living environments, lack of physical activity, social isolation—these are only some of the numerous factors leading to disease.

Physicians should inform their environment and the wider public on the importance of leading a healthy lifestyle. However, it is difficult to draw a line between well-intentioned advice and unsolicited intrusiveness. Consider the case of a patient who does not quit smoking despite suffering from obstructive lung disease. Is it appropriate for a physician to decline further care for the patient if he does not quit smoking? I offer the following response for consideration: the physician can express such intentions as a warning in the hope of achieving the desired effect; however, he should not carry out the threat and should continue with patient care.

12.3   Protection of Vulnerable Groups of Society

Preventive medicine is concerned less with individuals than it is with a wider population. Special attention is paid to vulnerable groups: children, the elderly and the lonely, drug addicts, disabled persons, immigrants and refugees, the poor, and workers in professions with health risks. High-quality drinkable water and a healthy diet are equally important. The task of preventive medicine is to analyze the factors that are detrimental to health, propose preventive measures, and inform the public.

Public statements can be unpleasant or even highly upsetting to the ruling elites. From the point of view of ethics, it is important that all working in preventive medicine keep their heads up high and do not yield to any attempts of silencing or whitewashing the conclusions. This is particularly true for those leading public health institutions whose duty is to encourage and enable impartial analyses and report about the most important issues relating to public health.

12.4   Epidemiology and Epidemiological Studies

One of the tasks of preventive medicine is the collection and analysis of data that are important for maintaining the accurate perceptions of health and disease in society at large.

Health-related data are for the most part collected anonymously in order to prevent us from infringing on citizens' privacy. In the chapter on ethical analysis, we reviewed the case of data collection for the Cancer Registry of the Republic of Slovenia. Since the treatment for most cancer patients is performed at several institutions, it is not possible to ensure full anonymity—otherwise, it would not be possible to prevent duplications or missing reports for each specific case. Building databases that include potentially identifiable personal information requires sound professional justification and legal grounds.

Surveys involving the general population or specific subpopulations of patients are normally anonymous. I would nevertheless advise caution when collecting sensitive personal information. A survey conducted via standard mail can serve as an orientation. A thorough analysis of professional stress, diet, lifestyle, or housing conditions, however, requires an in-depth conversation in person. For these cases, we must of course obtain informed consent in advance and ensure that the data will be anonymized prior to conducting any further analyses.

12.5   Return to the Introductory Parable

It would be misguided if activity in preventive medicine would be left entirely to professionals. How to prevent disease is a question for all physicians. Unfortunately, this is currently not the case: I have not yet heard that, for example, trauma specialists would be involved in a parliamentary discussion on legislative proposals pertaining to road safety.

A critical comment should also be addressed to our colleagues specializing in public health. All too often, their epidemiological conclusions are not followed by decisive preventive action. This is as if our introductory rescuer that headed to the source of the river would only stare at the dragon.

Suggested Reading

de Beaufort I. 'Please, sir, can I have some more?' Food, lifestyle, diets: respect and moral responsibility. Best Pract Res Clin Gastroenterol. 2014;28:235–45. https://doi.org/10.1016/j.bpg.2014.02.001.

Dubov A, Phung C. Nudges or mandates? The ethics of mandatory flu vaccination. Vaccine. 2015;33:2530–5. https://doi.org/10.1016/j.vaccine.2015.03.048.

Galanakis E, Jansen A, Lopalco PL, Giesecke J. Ethics of mandatory vaccination for healthcare workers. Euro Surveill. 2013;18:20627.

Grzybowski A, Patryn RK, Sak J, Zagaja A. Vaccination refusal. Autonomy and permitted coercion. Pathog Glob Health. 2017;111:200–5. https://doi.org/10.1080/20477724.2017.1322261.

Hirschberg I, Strech D, Marckmann G, editors. Ethics in public health and health policy: concepts, methods, case studies. Berlin: Springer; 2013. ISBN-13: 978-9400798052.

Scully IL, Swanson K, Green L, Jansen KU, Anderson AS. Anti-infective vaccination in the 21st century-new horizons for personal and public health. Curr Opin Microbiol. 2015;27:96–102. https://doi.org/10.1016/j.mib.2015.07.006.

ten Have H. Global bioethics: an introduction. Abingdon: Routledge; 2016. ISBN-13: 978-1138124103.

Ethics at the Beginning of Life

Abstract

Ethical and political debates on conception and abortion are often hot and one-sided. In an attempt to present a balanced view, we will discuss the whole spectrum of issues, from normal conception on one side to surrogate motherhood on the other. The main ethical issues are female/male imbalance regarding the side effects of contraception, uncritical use of the "morning-after pill," abortion on demand without accompanying counselling, preimplantation or prenatal screening for selection of fetus regarding gender, abortion for minor fetal abnormalities, artificial insemination for a healthy woman, anonymity in case of donation of sperm or egg cells, and financial compensation for surrogate motherhood.

World map with indication of gender ratio at birth. The average ratio between boys and girls is 1.06. The unusually high ratio in favour of boys is clear evidence of selective abortions, or of infanticides, of female offsprings (source: https://en.wikipedia.org/wiki/Human_sex_ratio)

© Springer Nature Switzerland AG 2019
M. Zwitter, *Medical Ethics in Clinical Practice*,
https://doi.org/10.1007/978-3-030-00719-5_13

Life obviously starts before birth. In this chapter, we will explore the ethical dilemmas pertaining to human activities before birth. We will gradually move from the conditions of normal fertility towards increasing medical intervention in what we term (to me very unpleasantly sounding) "human reproduction."

13.1   Natural Conception

13.1.1 Contraception

In principle, I do not see ethical objections to prevention of pregnancy. However, we have to point out that all forms of contraception for women, with the exception of the diaphragm (which, though, is not always effective), come with certain health risks. For men, the use of condoms is a perfectly safe and unproblematic method of contraception. From an ethical standpoint, the physician should emphasize that contraception must not be conceived as solely women's problem. On the contrary, we must encourage the use of contraception for men. Hormonal contraception is especially problematic in maturing girls: at a very young age, it interferes with the processes of bone development [1], affects hormonal balances, increases the risk for thromboembolic complications [2] and the risk for breast cancer in later years [3].

13.1.2 Morning-After Pill

The use of the morning-after pill, which prevents the implantation of the fertilized egg, is—if one is consistent—a form of abortion. Even if we for now disregard the objections regarding abortion (to be explored in the following section), the use of the morning-after pill is a severe intervention in women's hormonal state. Physicians must explain all health consequences and be clear that the morning-after pill must never become a regular method for the prevention of pregnancy.

13.1.3 Abortion

When it comes to abortion, many—and I am no exception—will find their opinions divided between clear-cut rejection and empathy for those who cannot find another way out from the distress of unwanted pregnancy.

In principle, it is clear: the embryo is already a unique human being. From the moment of fertilization, it is not possible to draw a clear line to separate the state of the embryo from that of the future human being. Any intervention interrupting this development is an intervention in life. From the ethical position, it does not matter whether abortion is performed surgically or with medication.[1] Abortion with

---

[1] Pills with mifepristone (progestorone blockers) and misoprostol (induction of uterine contraction) as active substances. In many countries, these drugs are sold as over-the-counter drugs.

medication can, in fact, turn out to be even more problematic: some women do not understand what they are doing when they take the three pills. Since abortions with medications are not captured in the healthcare statistics, the data showing a decrease in the number of abortions might be misleading.

A strong ethical viewpoint by steadfastly rejecting abortion would necessarily be uncompassionate and unfair to people in distress. Moralistic judgments would only deepen numerous difficult life situations. For many, life is anything but a fairytale: think of poverty, helplessness and the feeling of being lost, alcohol and drug abuse, psychological distress, lack of partner's compassion. It is our duty to understand that life circumstances are not always suitable for welcoming a new life.

Hence, there are problems on both extremes of the spectrum. A complete ban on abortion would leave no space for understanding and compassion towards women in distress. At the same time, such measures would foster illegal and often unprofessional practices that can severely harm women's health. However, to argue in favor of abortion by claiming that it is simply women's right, one might completely overlook the issue that abortion means terminating the life of a new human being. To reconcile the two viewpoints, we must do two things. Our first duty as individuals and as a democratic community is to guarantee the material and social conditions such that abortion as a result of societal pressures will no longer be needed. Our second duty is to introduce regular counselling as part of the legal abortion process. During counselling, the prospective parents should also hear about the possibility of putting the child up for adoption which, from an ethical viewpoint, is undoubtedly a lesser burden.

Thus far, we have only discussed abortion as a measure for terminating unwanted pregnancy. However, decisions for abortion are also made due to irregularities in fetal development or—even worse—due to the gender of the embryo. This brings us to prenatal diagnostics.

13.1.4 Prenatal Diagnostics

The main goal of all medical procedures during pregnancy is to determine whether the fetus is healthy. In case of irregularities in early fetal development, future parents can opt for abortion. Prenatal diagnostics therefore supports eugenics—preference for healthy children.

The wish of future parents for a healthy child is universal, and a decision for abortion in case of significant abnormalities is acceptable by most parents, physicians, and by society at large. It is necessary, however, to also respect those parents who reject abortion, and hence also prenatal diagnostics. "God blessed me with a child, and I want to keep this child even in the case of Down syndrome"; these parents earn our respect and deserve whatever help might be needed in case the child is not born healthy. In the case of parents' refusal of prenatal diagnostics, one should not limit the access to treatment for the child. A horrifying derivation of utilitarian ethics would be to advocate compulsory prenatal diagnostics, based on calculations showing that the costs of a 1000 ultrasound examinations are lower than the costs for treatment of a single child with Down syndrome.

In some parts of the world, statistics on birth ratios strongly point to selective abortions and female feticide or even female infanticide. For example, compared to the natural ratio of boys to girls at birth (1.03–1.07), the ratio in China is reported to be between 1.11 and 1.20 in favor of boys, based on different statistical estimates. Very similar estimates of skewed boy-to-girl birth ratios are known for India, Azerbaijan, Armenia, and Georgia [4]. It is estimated that every year at least 100,000 selective abortions are performed in India exclusively because the fetus is female [5]. In countries with strong prejudices about male superiority, it is advisable not to inform future parents about the gender of their child during prenatal diagnostics. However, in practice, such a recommendation is difficult to control.

13.1.5 Delivery at Home

A reasonable physician will not support a pre-planned delivery at home. In the first half of the twentieth century or earlier, delivery at home might have been safer than in overcrowded and often dirty public hospitals. Since then, the living conditions and hospitals have changed dramatically. But it is not only about hygiene—it is also about safety. While pregnancy and delivery are not a disease, a complication during home delivery may put at risk the health and the life of the mother and child. Idyllic ideas about a happy event within the warm surroundings of the family home may quickly turn into the reverse and may leave a permanent burden to the family. Thus, the autonomy of a pregnant woman has its limits, defined by responsibility for her own health and for the health of the baby.

13.2   Sterilization

Depending on national legislation, men or women above certain age may apply for sterilization. For men, the procedure (vasectomy) is simpler than for women (tubectomy or the blocking of Fallopian tubes). From the point of view of ethics, it is important that the person understands that after successful sterilization, fertility can be recovered only in exceptional cases. Sterilization without consent or even under coercion is illegal and represents a grave ethical offense.

13.3   Insemination or Egg Cell Donation

In the case of infertility, insemination with donor sperm or donation of egg cells are nowadays routine procedures. The main legal and ethical issues concern the status of the donor.

In my own country, Slovenia, and in many other countries, the donor remains anonymous. Only in cases of exceptional and justified health-related reasons can the court in a non-contentious procedure allow disclosure of the donor's identity. In several countries, however, donors are not guaranteed anonymity with the justification that the child upon coming out of age has the right to know who his or her biological parents are [6, 7]. I do not support such openness towards donor identity. First, the possibility of disclosure will deter many from donating sperm or egg cells. This may be an additional obstacle to the program already lacking donors. Abandoning anonymity upon the request of the child should not be one-sided: in this case, the donor should have a similar right to know who his or her biological children are. However, such openness represents a serious intervention in the family to which the child was born and can lead to abuse or blackmail.[2] My thoughts are therefore as follows: sperm or egg cell donation is an act performed from a wish to help those who cannot conceive a child in a natural way. This act confers neither responsibility nor any rights to the donor. The same holds for the other party: they received a gift and they retain all the responsibility. And the child? The child should rejoice in the fact that he or she is alive and should also accept the fact that his or her curiosity alone does not give the right to interfere with lives of other people.

13.4   Medically Assisted Insemination for Healthy Women

Let us consider briefly a heavily politicized topic: medically assisted fertilization for healthy women without a male partner. Is this question really worth the attention it is given? Is it possible and necessary to ban such a procedure?

In answering this question, we should try to let go of the ideological baggage. Following the method of ethical analysis, let us ask what, ethically speaking, are the costs and benefits for those involved in the problem: the future child, the future mother, the physician, and the wider society.

For the *child*, it is a choice between life and non-existence. Ethical evaluation will always prefer life over non-existence. There is therefore no doubt that medically assisted conception benefits the child.

In the case of the *future mother*—a single woman who wants to have a child—consideration of her decision becomes more complex. We must first clarify that by "single woman" we also refer to women living in a homosexual relationship; with respect to medically assisted conception and from an ethical standpoint, there is absolutely no difference between her and women living alone. There can be many possible reasons as to why a woman cannot find a male partner for starting a family: sexual orientation, unusual life experience and disappointments, unwillingness to accept the responsibilities and entrapment of family life, personality traits, disability, and mental disease. Will having a child really enrich her life or does it instead represent a burden? There are clear cases when the decision for having a child was not considered carefully enough and the woman would not be able to take care of the child. At the same time, it is undoubtedly true that motherhood brings happiness and meaningful life to many.

The physician is never only a medical expert, and respect for his own autonomy should also be considered. Some physicians may raise conscientious objection

---

[2] See also seminar paper nr. 42.

against the use of medically assisted fertilization. Those who participate in such a program have the right and responsibility to evaluate ethical costs and benefits in every individual case. To give a concrete example: the physician has the right and responsibility to refuse performing medically assisted conception in cases of severe mental disease when it is judged that the patient will not be able to offer a suitable environment for the child.

Wider society. A healthy single woman is not a patient; hence, the costs for medically assisted conception should not be covered by health insurance. It is reasonable to conclude, therefore, that the procedure should be self-funded.

Once we have sorted out the financial aspects of the procedure and established the physicians' right and responsibility to decline problematic applicants, it becomes clear that there are more prejudices than genuine problems concerning a few women who raise a child, either alone or with a female partner. In aging populations, every child counts. All parents, including single mothers, invest in child rearing. They spend their money mostly in the domestic economy; they do not spend it on luxuries or on holidays on the other side of the world.

Medically assisted conception for healthy women should not become a right; however, it is a possibility for enriching the lives of some women. There is no evidence to support arguments against such a possibility, on the grounds that a single woman or two women living together cannot offer a suitable environment to the child. Those who sincerely care for children's rights should turn to other, more pressing issues: they can help the hundreds, thousands of hungry, beaten, neglected, or simply unhappy children.

13.5   In Vitro Fertilization

In recent years, a stark increase in age at which the future parents decide to have children has been seen, especially in developed countries. Because fertility decreases with age, it is understandable that many parents require medical assistance in conception.

We will, of course, leave the professional and technical aspects of in vitro fertilization aside. However, it should be emphasized that such procedures are no longer uncommon. In Slovenia with a population of 2.1 million, there will soon be 10,000 children whose life started "in vitro." In other words, on average one pupil in every classroom.

The induction of ovulation and in vitro fertilization of egg cells leads to up to a dozen fertilized eggs—zygotes, of which usually one or two are implanted. The remaining zygotes are deeply frozen: some might be used in a later repetition of the procedure; others will have to be discarded. The sheer fact that unused zygotes must often be discarded has led some to categorically reject the entire program of in vitro fertilization. In principle, they are right: the fusion of egg and sperm cells marks the start of a new life. However, I am strongly convinced that such a narrow viewpoint is inappropriate. Does this mean that we must for principled reasons sacrifice the happiness of thousands of people for whom such medical procedures help towards a rich family life?

The procedures of in vitro fertilization are also used in cases when we need to select among the fetuses to avoid genetic diseases. Ethical considerations here are similar to those in prenatal diagnostics: medicine supports the wishes of the parents to have only healthy children. Obviously, in vitro fertilization is ethically not acceptable when it serves to choose the gender or physical or mental characteristics of the child.

13.6 Surrogate Motherhood

The most significant deviation from natural conception and pregnancy is surrogate motherhood. Surrogacy arrangement means that the fetus is implanted into the womb of another woman who gives birth to the child and then gives the child to the "biological parents."

Surrogacy opens the door to severe abuse. The indication for surrogacy is often not a medical one, that is, the inability of the biological mother to carry a pregnancy and give birth, but rather the biological mother wants to avoid all the risks and discomforts of pregnancy and birth. In some parts of the world, for example, India, the so-called *rent-a-womb* is already a well-established source of income for poor women. This practice is illegal in many countries and is ethically very problematic. Few healthy women would accept the role of a surrogate mother purely for altruistic motives, and in the large majority of cases, it is the money that drives the decision. Surrogate motherhood is a clear case of disrespect of the ethical principle of justice. Ethically, it is hard to defend exploitation of the poor, who are forced to put their health and well-being at risk.

The second problematic issue is that pregnancy is not merely a waiting time, but rather a precious period for both parents to prepare for their new role. Talking, or singing to a child before birth, is a most common and enriching experience. Likewise, the father also gradually adapts to his new role. Thus, all events during pregnancy prepare the family for the great change in family life and for the (often difficult) period of caring for a newborn baby. This role of pregnancy as preparation for parenthood is missing if the only reminder of the new family member is a marked date on a calendar.

Finally, we must also think of some unfortunate potential scenarios. What happens if the newborn child is not healthy? Would the "biological parents" return him or her as we do with malfunctioning washing machines in stores? If the surrogate mother during pregnancy or delivery gets a disease or even dies—who is responsible and who takes care of her other children? Or to take the case of the wealthy pediatrician Dr. Elizabeth Stern from New York: what happens when the surrogate mother does not want to hand over the child and cancels the contract? To whom does the child belong [8]?

Clearly, there are exceptions here as well. One such case is discussed in the first seminar. I invite the reader to judge for him or herself if there are ethically acceptable forms of surrogacy.

References

1. Ziglar S, Hunter TS. The effect of hormonal oral contraception on acquisition of peak bone mineral density of adolescents and young women. J Pharm Pract. 2012;25:331–40.
2. O'Brien SH. Contraception-related venous thromboembolism in adolescents. Semin Thromb Hemost. 2014;40:66–71.
3. Primic-Žakelj M, Evstifeeva T, Ravnihar B, Boyle P. Breast-cancer risk and oral contraceptive use in Slovenian women aged 25 to 54. Int J Cancer. 1995;62:414–20.
4. https://en.wikipedia.org/wiki/Sex-selective_abortion
5. MacPherson Y. Images and icons: harnessing the power of media to reduce sex-selective abortion in India. Gend Dev. 2007;15:413–23.
6. http://www.theguardian.com/commentisfree/2015/aug/18/anonymous-sperm-donation-is-flawed-just-ask-donor-conceived-children
7. https://en.wikipedia.org/wiki/Sperm_donation_laws_by_country
8. https://en.wikipedia.org/wiki/Baby_M

Suggested Reading

Bawany MH, Padela AI. Hymenoplasty and Muslim patients: Islamic ethico-legal perspectives. J Sex Med. 2017;14:1003–10. https://doi.org/10.1016/j.jsxm.2017.06.005.

Benward J. Mandatory counseling for gamete donation recipients: ethical dilemmas. Fertil Steril. 2015;104:507–12. https://doi.org/10.1016/j.fertnstert.2015.07.1154.

Birkhäuser M. Ethical issues in human reproduction: protestant perspectives in the light of European protestant and reformed churches. Gynecol Endocrinol. 2013;29:955–9. https://doi.org/10.3109/09513590.2013.825716.

Blyth E, Crawshaw M, Frith L, Jones C. Donor-conceived people's views and experiences of their genetic origins: a critical analysis of the research evidence. J Law Med. 2012;19:769–89.

Burrell C, Edozien LC. Surrogacy in modern obstetric practice. Semin Fetal Neonatal Med. 2014;19:272–8. https://doi.org/10.1016/j.siny.2014.08.004.

Brezina PR, Kutteh WH. Clinical applications of preimplantation genetic testing. BMJ. 2015;350:g7611. https://doi.org/10.1136/bmj.g7611.

Catlin AJ, Volat D. When the fetus is alive but the mother is not: critical care somatic support as an accepted model of care in the twenty-first century? Crit Care Nurs Clin North Am. 2009;21:267–76. https://doi.org/10.1016/j.ccell.2009.01.004.

Gong D, Liu YL, Zheng Z, Tian YF, Li Z. An overview on ethical issues about sperm donation. Asian J Androl. 2009;11:645–52. https://doi.org/10.1038/aja.2009.61.

Holten L, de Miranda E. Women's motivations for having unassisted childbirth or high-risk homebirth: an exploration of the literature on 'birthing outside the system'. Midwifery. 2016;38:55–62. https://doi.org/10.1016/j.midw.2016.03.010.

Howe D. Ethics of prenatal ultrasound. Best Pract Res Clin Obstet Gynaecol. 2014;28:443–51. https://doi.org/10.1016/j.bpobgyn.2013.10.005.

Johnson BR Jr, Kismödi E, Dragoman MV, Temmerman M. Conscientious objection to provision of legal abortion care. Int J Gynaecol Obstet. 2013;123(Suppl 3):S60–2. https://doi.org/10.1016/S0020-7292(13)60004-1.

Kamm FM. Bioethical prescriptions: to create, end, choose, and improve lives. Oxford: Oxford University Press; 2013. ISBN-13: 978-0190649616.

Lackie E, Fairchild A. The birth control pill, thromboembolic disease, science and the media: a historical review of the relationship. Contraception. 2016;94:295–302. https://doi.org/10.1016/j.contraception.2016.06.009.

Lanzone A. Ethical issues in human reproduction: catholic perspectives. Gynecol Endocrinol. 2013;29:953–4. https://doi.org/10.3109/09513590.2013.825717.

Londra L, Wallach E, Zhao Y. Assisted reproduction: ethical and legal issues. Semin Fetal Neonatal Med. 2014;19:264–71. https://doi.org/10.1016/j.siny.2014.07.003.

Mercurio MR. Pediatric obstetrical ethics: medical decision-making by, with, and for pregnant early adolescents. Semin Perinatol. 2016;40:237–46. https://doi.org/10.1053/j.semperi.2015.12.013.

Minear MA, Alessi S, Allyse M, Michie M, Chandrasekharan S. Noninvasive prenatal genetic testing: current and emerging ethical, legal, and social issues. Annu Rev Genomics Hum Genet. 2015;16:369–98. https://doi.org/10.1146/annurev-genom-090314-050000.

Morrell KM, Chavkin W. Conscientious objection to abortion and reproductive healthcare: a review of recent literature and implications for adolescents. Curr Opin Obstet Gynecol. 2015;27:333–8. https://doi.org/10.1097/GCO.0000000000000196.

Neri M, Turillazzi E, Pascale N, Riezzo I, Pomara C. Egg production and donation: a new frontier in the global landscape of cross-border reproductive care: ethical concerns. Curr Pharm Biotechnol. 2016;17:316–20.

Pace TN. Bioethics: issues and dilemmas. Waltham, MA: Nova Biomedical; 2013. ISBN-13: 978-1617282904.

Ravelingien A, Provoost V, Pennings G. Open-identity sperm donation: how does offering donor-identifying information relate to donor-conceived offspring's wishes and needs? J Bioeth Inq. 2015;12:503–9. https://doi.org/10.1007/s11673-014-9550-3.

Samorinha C, Pereira M, Machado H, Figueiredo B, Silva S. Factors associated with the donation and non-donation of embryos for research: a systematic review. Hum Reprod Update. 2014;20:641–55. https://doi.org/10.1093/humupd/dmu026.

Schenker JG. Human reproduction: Jewish perspectives. Gynecol Endocrinol. 2013;29:945–8. https://doi.org/10.3109/09513590.2013.825715.

Serour GI, Serour AG. Ethical issues in infertility. Best Pract Res Clin Obstet Gynaecol. 2017;43:21–31. https://doi.org/10.1016/j.bpobgyn.2017.02.008.

Zampas C. Legal and ethical standards for protecting women's human rights and the practice of conscientious objection in reproductive healthcare settings. Int J Gynaecol Obstet. 2013;123(Suppl 3):S63–5. https://doi.org/10.1016/S0020-7292(13)60005-3.

Pediatrics

14

Abstract

The chapter starts with the ethical issues related to a severely handicapped new-born, a condition that often requires a series of medical and surgical interventions. When chances for long-term survival and life without continuous significant medical support are slim or non-existent, the attitude of the parents towards intensive medicine or palliative care should be considered. In most other situations in pediatrics, however, the ethical validity of decisions made by parents or caregivers is not a question of autonomy of surrogate decision-makers but should be based on the principle of beneficence towards the child. Even if the autonomy of a child is limited, it should be respected to the highest reasonable level, and any medical intervention should be accompanied with an appropriate explanation. The chapter includes discussion of some topics specific to the young: mandatory vaccination, suspected child abuse, genetic testing in childhood, reactions to the death of a child, and clinical research in a pediatric population.

In the family of my great-grandmother, two boys died before the age of 2. In the past, families were large, and the death of a child was a common experience. Nowadays, the death of a child is considered a tragedy. Since medicine is considered to be all-powerful, physicians are often accused of malpractice

A word on ethics in pediatrics is a word on adhering to the principle of beneficence, but at the same time a word on the limited autonomy of children. We move in a triangle: the child, the child's parents or guardians, and the physician with the support of other healthcare and non-healthcare workers. All of the involved have to put the interest of the child first.

14.1   The Newborn

The main ethical questions pertain to the medical care of severely handicapped children. Possible causes are diverse: hereditary diseases, inborn abnormalities, infections, and other harmful events during pregnancy, premature birth (especially before the 24th week of pregnancy), complications during labor, and complications immediately after childbirth.

We know a great deal about Down syndrome: the physician's main duty is to help the parents overcome the initial shock and to present the sources of help the family will need. Similarly, the management of minor developmental abnormalities, such as cleft palate, is well understood and requires no further elaboration on the ethical perspective. Ethically uncontroversial are also those cases where adequate care can help the child to overcome the complications and live a healthy life. A much greater issue, however, is the treatment of severely ill children. In the first days, perhaps even weeks, the medical team will do everything to save the child. Even if the chances look slim, it is, of course, worth trying; at the same time, the parents gain time to recover from the initial shock. If the child's condition does not improve, but in fact stabilizes or even deteriorates, the medical team has to first hold an internal meeting and afterwards have a consultation with parents. In decisions about intensive treatment, such as major surgical procedures in a newborn with several inborn abnormalities, the physicians will also take into account the parents' opinion. In cases when life without dependence on intensive care support cannot be expected, and parents are not in favor of aggressive treatment, the decision of the medical team to limit treatment to palliative care is acceptable.

In most developed countries, newborn screening programs for infants have been implemented with the aim of the early detection of conditions that are treatable but not clinically evident. The programs include blood screening tests, pulse oximetry for detection of congenital heart defects, and tests for vision and hearing loss. While these examinations represent a minor burden for the child, early discovery of a disease can provide an enormous benefit for the child's future development. In countries in which these programs are mandatory, forgoing such examinations can therefore rightly be regarded as an unethical act of neglecting the child. Parents planning delivery at home have to be reminded that their newborn should be examined by a neonatologist.

14.2 Vaccination

Vaccination has a three-fold purpose. The first is to protect an individual against a contagious disease, the second is to prevent the spread of the contagious disease, and the third is to protect those who cannot be vaccinated due to disease or conditions of immunodeficiency. From the perspective of public health, it is perfectly clear—vaccination against contagious disease has notably contributed to a significant increase in life expectancy. Preventive medicine, which includes programs of mandatory vaccination, and improved living, working, and hygienic conditions have contributed more to increased life expectancy than all other branches of medicine combined.

We will not go into a discussion on the benefits of vaccination. Instead, we will focus on communication with parents who oppose vaccination.

In the absence of a disease or another condition due to which vaccination would be contra-indicated, abandoning vaccination may be harmful to the child. The risk associated with vaccination is minimal, and the reports on harmful side effects of vaccination are exaggerated or have even proven to be fraudulent.[1] At the same time, it is a fact that many diseases considered nearly eradicated are reoccurring: measles, tuberculosis, even poliomyelitis. Consequences of infection with these diseases in adults who have not been vaccinated in childhood are often severe. Similar holds for vaccination against the human papillomaviruses, which are a significant factor in head and neck cancer and the main cause of cervical cancer—a disease that kills between 40 and 50 women per year in Slovenia alone.

While all scientific evidence points to the benefits of vaccination, some parents oppose vaccination on the basis of utterly selfish reasoning. "If all other children are vaccinated, the risk for my baby is low; therefore, I do not agree with vaccination"—these words are rarely heard, yet often an important argument of the parents.

To be better positioned against the groups who argue for the "freedom of choice," physicians must actively engage in public discussions. It is our duty to lay out all the arguments to the parents: the principle of justice in protecting the entire population, the potential harm done to their child, and increased risk for those children that cannot be vaccinated.[2] When parents still do not agree, they may be asked to sign a document that they reject vaccination despite explanation. In several countries, proof of vaccination is required in order for the child to be admitted to kindergarten or public school.

How to proceed in cases if the child gets a disease that could have been prevented with vaccination? Should we agree with the proposal that in such cases parents cover the costs of treatment? Personally, I am conflicted about such situations. On the one hand, it is true that the costs of treatment can be very high and that a child's treatment must not depend on the financial means or whims of the parents. On the other hand, we know that it is mostly the parents from higher classes of society that reject vaccination. If we follow the principle of justice, then we will support the possibility that the well-off parents cover the costs of their capricious behavior themselves. There is definitely also the possibility to take the parents to court for child neglect. The complaint would be even more justified if the child who was not vaccinated due to parents' opposition would infect another child with immunodeficiency for whom the disease can be severe or even fatal.

---

[1] In 1998, Andrew Wakefield published an article in The Lancet in which he supposedly proved that vaccination against measles, mumps, and rubella can be the cause of autism and colitis. It was later confirmed that the data in the article were forged. Dr. Wakefield's medical license in Great Britain was revoked.

[2] Recently, Dr. Urh Grošelj presented an interesting comparison that can be used when trying to convince others why vaccination in healthy children is justified. Everyone agrees that it is safer for the baby in a car to be fastened in the child safety seat. Even if once in every 10,000 accidents it happens that a child is severely injured due to the safety belt, we will not oppose the use of child safety seat on the basis of this single case. Similar holds for vaccination: it is unreasonable to reject a measure, which is undoubtedly beneficial to the majority of children, due to very rare instances of complications.

The latter remark applies to those rare cases when complications are very likely caused by vaccination. These cases must be registered, carefully monitored, and analyzed. Immediately and without any legal proceedings, the family should be offered the right to compensation.

14.3 Child Neglect and Maltreatment

The list alone of the potentially serious threats to healthy development in children is lengthy: poverty, hunger, poor living conditions, war, physical threats by parents or guardians, psychological abuse, parents' excessive ambitions, sexual abuse, mental distress due to fights between parents, divorce and fights over guardianship, alcohol and other drugs in family, very unbalanced diet due to parents' preconceptions, and rejection of urgent treatment in cases of serious disease.

As with other medical duties, the physician's primary responsibility is to establish a diagnosis. This is not always easy because parents will frequently conceal the true causes; they will, for example, claim that the child fell or had an accident. Equally difficult is to take action, which must, of course, be beneficial to the child. In a society of excessive protection of privacy and noninvolvement in the lives of others, taking measures against parents or guardians is challenging and unfortunately often successful only when the consequences for the child are already profound.

We must applaud the courage of those physicians who firmly side with the child, often in the face of threats of violence against them. At the same time, we must protect a physician who suspected child abuse and later found out that there were no grounds for his suspicion. Everyone—parents, fellow physicians, and the public—must understand that suspicion is not the same as an allegation. Any form of contempt or shaming of the physician who "committed" groundless suspicion is unacceptable; with the physician's apologies, the matter should be settled.

In addition to meat and fish, which are excluded from a vegetarian diet, a vegan diet excludes milk, dairy products, and eggs. To put it mildly, a vegan diet for children is highly problematic. To prevent severe and mostly irreversible dysfunction in the child's physical and mental development, parents must specifically add to the otherwise very unbalanced diet vitamins B12 and D, riboflavin, calcium, iron, zinc, other minerals and essential amino acids, and omega-3 fatty acids [1]. This fact alone proves that a vegan diet represents unnatural nutritional restriction to which human physiology is not adapted. An important open question is whether parents who obstinately cling to their different views on "natural living" will, in fact, follow the recommendations and add all the above-mentioned "chemical, artificial" supplements to their child's vegan diet. When faced with vegan parents, the physician should be a resolute advocate of the child's benefits and should clearly warn the parents of all risks. At the same time, the physician should carefully monitor the child so as to act upon the first signs of dysfunctions in his/her development.

One form of child neglect is addiction to video games and internet communication [2]. Many parents, especially younger ones (often having the addiction themselves) are not aware just what the consequences are for the mental development of a child who spends most of his time and attention in the virtual world. Autistic children are two times more frequently addicted to the virtual world of video games and internet communication; it remains unclear, however, what is the cause and what is the effect of this pattern. Can autism be the consequence of enclosing oneself in the virtual world, or do autistic children simply find their place on the screen, rather than in the frequently traumatic experience of the real world?

14.4   Chronic Disease and Communication

We often say that a serious disease of an individual represents a burden for the whole family. This statement is especially true when we speak about the serious disease of a child. The child and his disease are at the epicenter of all family activities. While such careful attention is, on the one hand, human, laudable, and touching, there are downsides to it as well. The sister of the young patient is left to herself; the parents will intervene in every (otherwise perfectly normal) row or fight between the kids. Their protective intervention will prevent the development of a normal relationship between the children.

A child's serious disease places an immense emotional burden on all involved. Although the physician might be used to it, he must show compassion in communication with children and parents. It is almost always inappropriate if we unduly try to establish whose fault it was or how delayed the diagnosis was: if a specialist wants to communicate to his colleague at the primary level that an accurate diagnosis might have been reached earlier, he should do so in direct contact, and not through parents.

Empathy goes hand in hand with mutual trust. On the physician's part, this also includes an appropriate degree of humility. The key enemies of trust are a physician's arrogance, an air of superiority, and omniscience. No matter how annoying, an experienced physician patiently listens to the parents who found advice on the web and helps them in obtaining a second, independent opinion. He is aware that all this thirst for additional information is only a sign of the great energy that parents devote to their child. If the physician does not harness this energy, it may turn against him.

Already in his early age, a child with chronic disease can develop a very good understanding of numerous medical procedures. Children are therefore rightly upset if they are ignored when we are giving an explanation or even trying to trick them. The importance of partnership in the physician–patient relationship is even more crucial in pediatrics: the child's understanding, trust, commitment, and adherence in performing his responsibilities and also his gratitude often surpass those of adults.

14.5   Genetics

In examining the genetic nature of a disease, we are witnessing unprecedented progress, which, however, also brings along significant ethical dilemmas. Genetics is, therefore, the topic of the next chapter.

In this chapter on pediatrics, we wish to focus on questions related to surrogate decision-making, which should adhere to the ethical principle of beneficence. Parents and guardians might require genetic testing even in cases when it is not benefiting the child. To prevent or limit these risks, we must be consistent in adhering to the rule that underage patients should only be tested in cases in which a medical measure or treatment would be indicated in childhood.

14.6   Death of a Child

In village cemeteries, one can often still see small rather overgrown graves. Yes, in the past child death was quite common. Families had many children, and the death of a child was not considered a tragedy for the family. Since, nowadays, any case of child death is considered unnatural, and because medicine is viewed as omniscient and omnipotent, the physician is the obvious candidate to be guilty of an unfortunate event. At the same time, parents or guardians seek ways, willingly or not, to offload their part of the responsibility for a traffic accident or for a late consultation with the physician.

Remarks about communication with chronic patients apply even more for cases of child death: empathy, patience, humility. In response to potential accusations, the physician should offer a detailed and clear explanation. When facing serious accusations, the physician will ask a colleague or head of the department for help. In cases of overt malpractice (which is fortunately quite rare), and in cases of suboptimal diagnostics and treatment due to circumstances that were beyond the responsibility of a particular physician, eventual financial compensation may not be of major concern. What parents wish to hear is a clear statement that physicians will do everything to prevent similar events from reoccurring in the future.

14.7   Research

I definitely agree that it would be highly inappropriate if we approached clinical trials rashly. A thorough consideration of all existing knowledge, a clear hypothesis, objectives of a trial, precise description of diagnostic and therapeutic procedures, and statistical planning are essential elements of every protocol, regardless of whether one is dealing with adult patients or children.

All documents on ethics consider children to be a vulnerable group. Yet, we must not forget that there is no progress without research. When it comes to children, the main problem is lack of commercial interest for clinical trials in pediatric populations.[3] Let us illustrate this with the situation in pediatric oncology. The first truly astonishing results in systemic cancer treatment were achieved in the

---

[3] In 2015, Nuffield Council on Bioethics published recommendations on participation of children in clinical research: http://nuffieldbioethics.org/wp-content/uploads/Children-and-clinical-research-full-report.pdf.

treatment of childhood cancers. Half a century ago, pediatric oncology was paving the way for all other branches of oncology. Since then, not much has changed for a child with cancer, and the majority of drugs in pediatric oncology have been in use for decades. This is in sharp contrast to the situation with common cancers in adulthood, which are in the mainstream of clinical research, and where each year brings a dozen new drugs.

References

1. Mesina V, Mangels AR. Considerations in planning vegan diets: children. J Am Diet Assoc. 2001;10:661–9.
2. http://www.techaddiction.ca/child-video-game-addiction.html

Suggested Reading

Botkin JR. Ethical issues in pediatric genetic testing and screening. Curr Opin Pediatr. 2016;28:700–4.

Chervenak FA, McCullough LB. Healthcare justice and human rights in perinatal medicine. Semin Perinatol. 2016;40:234–6. https://doi.org/10.1053/j.semperi.2015.12.011.

Chervenak FA, McCullough LB. Ethics in perinatal medicine: a global perspective. Semin Fetal Neonatal Med. 2015;20:364–7. https://doi.org/10.1016/j.siny.2015.05.003.

Grootens-Wiegers P, Hein IM, van den Broek JM, de Vries MC. Medical decision-making in children and adolescents: developmental and neuroscientific aspects. BMC Pediatr. 2017;17:120. https://doi.org/10.1186/s12887-017-0869-x.

Hein IM, De Vries MC, Troost PW, Meynen G, Van Goudoever JB, Lindauer RJ. Informed consent instead of assent is appropriate in children from the age of twelve: policy implications of new findings on children's competence to consent to clinical research. BMC Med Ethics. 2015;16:76. https://doi.org/10.1186/s12910-015-0067-z.

Hoytema van Konijnenburg EM, Teeuw AH, Ploem MC. Data research on child abuse and neglect without informed consent? Balancing interests under Dutch law. Eur J Pediatr. 2015;174:1573–8. https://doi.org/10.1007/s00431-015-2649-7.

Orzalesi M, Danhaive O. Ethical problems with neonatal screening. Ann Ist Super Sanita. 2009;45:325–30. Review. PMID: 19861738.

Parasidis E, Opel DJ. Parental refusal of childhood vaccines and medical neglect laws. Am J Public Health. 2017;107:68–71.

Rapoport A, Morrison W. No child is an island: ethical considerations in end-of-life care for children and their families. Curr Opin Support Palliat Care. 2016;10:196–200. https://doi.org/10.1097/SPC.0000000000000226.

Rosenberg AR, Starks H, Unguru Y, Feudtner C, Diekema D. Truth telling in the setting of cultural differences and incurable pediatric illness: a review. JAMA Pediatr. 2017;171:1113–9. https://doi.org/10.1001/jamapediatrics.2017.2568.

Schilling S, Christian CW. Child physical abuse and neglect. Child Adolesc Psychiatr Clin N Am. 2014;23:309–319, ix. https://doi.org/10.1016/j.chc.2014.01.001.

Genetics

15

Abstract

Three characteristics make testing for genetic abnormalities different from all other medical diagnostics: life-long validity for the tested person; implications for a broader circle of blood-related family members; and the possible emergence of genetic information that was not solicited but may nevertheless influence the life of the person and his/her family. For these reasons, professional counselling prior to testing is an essential requirement. In prenatal diagnostics, genetic testing is ethically acceptable for the elimination of embryos with genetically determined diseases, but not for selection regarding gender or personal characteristics. For children, testing for adult-onset diseases should not be performed, and genetic testing should be reserved for conditions for which an immediate intervention may benefit the child. Great caution and consent prior to testing are recommended in dealing with unsolicited information. Unsolicited information regarding paternity should not be communicated, with the exception of clear proof of incest involving a minor.

© Springer Nature Switzerland AG 2019
M. Zwitter, *Medical Ethics in Clinical Practice*,
https://doi.org/10.1007/978-3-030-00719-5_15

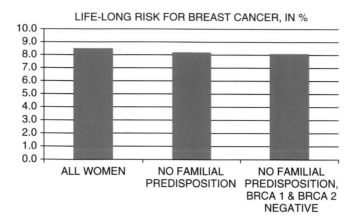

Common misunderstanding: a negative genetic test indicates that your risk for cancer is low. Since most cancers are not linked to a detectable mutation, a negative result of known mutations has virtually no effect upon the risk for a particular cancer (source of data for the graph: Cancer Registry of Slovenia). Commercial testing without counselling is of very limited benefit and may be misleading

Genetic testing is different from other medical diagnostics in two crucial aspects. First, findings in other types of diagnostics relate to a particular time, whereas the results of genetic testing hold for life. Second, findings of genetic tests may affect the lives of other blood relatives. Due to these two distinctive characteristics, consultancy is an essential element of genetic diagnostics, both before deciding on testing and after the results are known.

In this chapter, we will discuss only those genetic characteristics that encompass the entire body and are passed from one generation to the next. We are, therefore, leaving aside gene changes that occur de novo during the course of a lifetime and affect only a part of the organism. For example, all cancer diseases are due to mutations in the genetic code; however, such mutations (with rare exceptions) are not inheritable.

We begin this chapter by describing the main areas that are influenced by genetic testing. In the second part, we will focus on the ethical evaluation of certain procedures in genetic technology and knowledge.

15.1   What Information Can Be Gained from Genetic Testing?

Genetic tests can be performed on embryos (prior to implantation or during intrauterine life) or on humans after birth and may be done on individuals or on wider populations.

Due to the rapid development of laboratory methods and new knowledge, it would be ill-considered to attempt to provide a definite list of all areas influenced by genetics. Having provided this disclaimer, let us nevertheless try and show the range of findings that can be obtained by genetic tests.

- Preimplantation and intrauterine diagnostics: gender, presence/absence of specific hereditary disorders, HLA typing, physical and mental characteristics of the future person.
- Diagnostics in pediatrics and in adults: certain metabolic disorders, diagnosis or confirmation of hereditary diseases.
- Heredity as a cause of some diseases, such as certain forms of cancer, which often arise also spontaneously.
- Genetic predispositions towards certain diseases.
- Metabolism, pharmacokinetics, prediction of tolerance for and efficacy of drugs.
- Confirmation or exclusion of parenthood, usual fatherhood.
- Parenthood between close blood relatives (incest).

The differences between genetic testing and other medical diagnostics were discussed in the introductory paragraph. At this point, we wish to highlight another characteristic of genetics. In diagnostic examinations, we normally start with a straightforward question, and we expect a clear answer in return. However, in genetic testing, we often obtain information that we have not sought. Modern technology may reach beyond individual genes and perform analysis of a broad list of genes or even of the whole human genome. Even if the medical question concerned only a single gene, it may be easier to perform a broader analysis.[1] Apart from the desired information, such an analysis can reveal information about predispositions for other hereditary disorders and about genetic variants whose meaning we do not yet understand.

Gene analysis can also provide evidence of incest or disprove alleged fatherhood.

15.2   Disease Exclusion, Choice of Gender, and Selection of Other Traits of the Future Child

For over the last 70 years, genetics has carried the stains of racism driven by Nazi and fascist ideologies, xenophobia, and eugenics programs. It is, therefore, necessary to clearly distinguish between the understandable desire to be healthy, which includes the desire for healthy offspring, and unacceptable efforts towards unnatural improvements of genetic characteristics of future generations.

In the process of in vitro insemination, it is possible to perform preimplantation genetic diagnostics. Preimplantation genetic diagnostics is ethically justified in a family with a history of the hereditary disease and with a significant chance that a serious, life-threatening disease can be transferred to the next generation. Examples of such diseases include cystic fibrosis, sickle-cell disease, Tay-Sachs disease, Duchenne muscular dystrophy, beta-thalassemia, or Huntington disease. In some countries, preimplantation genetic diagnostics is also allowed in women above

---

[1]The situation is similar to the usual blood test: it is easier for a physician and for a laboratory technician to order and perform the analysis of the full range of hematological and biochemical parameters than to select only a few specific tests.

35 years of age, who have a higher risk of giving birth to a child with Down syndrome or other inborn disorders.

Performing preimplantation genetic diagnosis and embryo selection on the basis of the presence of other mutations, which potentially increase the risk of disease development in adulthood is ethically controversial. An example of a controversial case is the potential presence of the BRCA1 or BRCA2 mutation, both of which increase the risk for (in most cases curable) breast cancer. We must be aware that in the future, an increasing number of links between the genetic code and predisposition to specific diseases will be identified. Our lives and the lives of our descendants simply cannot be fully designed in advance.

Until recently, selective implantation of female-only embryos was frequently performed in families with a disease linked to the X chromosome (examples include hemophilia and Duchenne muscular dystrophy). Nowadays, such measures are no longer needed; precise gene analyses allow for the exclusion of all embryos with the disease-carrying mutation. It is, therefore, possible to exclude either female embryos that would be carriers of the disease as well as male embryos that would be affected by the disease. There are therefore no medical indications for gender-based selection of implanted embryos. Any other non-medical reasons for gender-based embryo selection are ethically unacceptable. If there are already four sons or four daughters in a family, this is not an ethically acceptable reason to perform a medical procedure that would guarantee the desired gender of the fifth child.

Ethically problematic is embryo selection based on transplantation antigens. A child with a heritable disease such as beta-thalassemia can be cured with bone marrow transplantation. If there is no other donor among the relatives who would match the transplantation antigens of the sick child, the parents may decide for another child. In such a case, in vitro insemination is performed, choosing an embryo compatible with the future recipient. At birth, immature hematopoietic cells are obtained from the cord blood. The donor suffers absolutely no harm during the procedure and will continue to live normally.

Is embryo selection for the purpose of providing "spare parts" for the brother ethical? If we approach the question with the method of ethical analysis, we find that no one is harmed in the procedure: the complete treatment with cord blood cell transplantation brings health to the child with thalassemia. For the newborn child, life is also more valuable than non-existence, no matter how unusual the motivation for his or her birth was. Nevertheless, we must remain very restrained in our ethical evaluation of the embryo selection procedures for providing the "spare parts": we cannot compare the use of cord blood cells, which would otherwise be discarded, with the transplantation of other organs or tissues, which is ethically absolutely unacceptable.

Fortunately, each person is unique. People are not like cars coming off an assembly line. All differences and minor deviations, for example, short stature or color blindness,[2] simply cannot and should not be eliminated by medical interventions.

---

[2] In Europe, approximately 8% of males have color blindess and cannot distinguish between red and green. Color blindness is caused by a mutation on the X chromosome, which is why it is more frequent in men.

It would, therefore, be unethical if we used the preimplantation diagnosis to select embryos based on physical or mental characteristics. When there are no clear medical indications, preimplantation genetic diagnosis is medically and ethically unjustified.

15.3 Genetic Testing in Childhood and Adulthood

Genetic testing is not an exception regarding the general rule of medical diagnostics: an examination is justified if its outcomes are informative for future action. Let us add another general rule: along with *the right to know*, we must also respect *the right not to know*.

Genetic tests in children are justified when such examinations are helpful in establishing a diagnosis and in choosing the most appropriate treatment. In healthy children, genetic testing is justified only in cases when appropriate medical measures or counselling (e.g., a modified diet) would alleviate, postpone, or eliminate signs of disease. However, children should not be tested for diseases that develop in adulthood.

Some adult diseases are linked to a genetic disposition. Heredity as a predisposition to cancer is suspected when several blood relatives develop the same type of cancer or when cancer develops at an unusually early age. Consider breast cancer as an example: in approximately 5% of the patients, BRCA-1 or BRCA-2 gene mutations significantly influence cancer development. From the patient's perspective, it is important to identify such a genetic predisposition and remain alert to the risk of cancer development in the other breast. These patients also significantly more often develop ovarian cancer. However, despite the advantage that such information brings to the patient, we must never insist on genetic analysis. The patient must be informed about this possibility, but also about the positive and negative consequences of learning about the potential genetic underpinnings of the disease. Tests can therefore only be performed if the patient is thoroughly informed and gives full consent. When receiving the results, the exam should never be sent by mail; rather, the patient has to be invited for additional conversation and consultancy.

If the genetic nature of cancer is confirmed, this is considered a warning to all other blood relatives. The patient decides about whom to inform and invite for potential testing. As noted above, genetic testing for diseases that develop in adulthood is never performed in underage patients—information gained would only serve to confuse and burden them psychologically. It follows that we never inform underage patients about the results of genetic tests on their relatives, who are also advised against sharing the results widely.

Full information and consent of a patient or a healthy person prior to testing and counselling once the results are known are therefore necessary components of genetic testing. Informing and counselling require in-depth specialist knowledge in the field of genetics and in the medical speciality, for example, oncology. At the same time, providing counselling requires excellent communication skills and even readiness to relieve unexpected human distress. While the examination itself can be performed by a machine, communication must by no means be limited solely to written information.

Limitation of testing to selected indications and the need for personal communication of information, consent prior to the examination, and counselling are strong arguments against entrusting genetic examinations to private companies driven solely by their financial interests. In our understanding, it is entirely unacceptable that people order their genetic examinations on the web. Although certain laboratories in the USA offer analyses of the whole genome for a price of 1000 dollars, such activities are most often of no benefit or even harmful.

Should gene testing be performed in healthy adults if blood relatives do not have a specific disease? The answer is a very resolute "no." Many lay people falsely believe that genetics can predict a disease. Consider again breast cancer as an example. When there are no cases of the disease in the family, and there is, therefore, no reasonable suspicion of a BRCA gene mutation, a negative result of genetic testing is absolutely useless and can even be misleading. The woman has still almost equal chances of developing breast cancer spontaneously (i.e., without a genetic predisposition). A negative result of genetic testing may lead to an unjustified feeling of security. Nevertheless, she should continue with regular participation in programs for early breast cancer detection.

15.4   Accidental Findings

Genetic testing can reveal unexpected information:

- Genetic abnormalities that point to another, currently clinically silent (asymptomatic) disease
- Genetic abnormalities or variants of uncertain or unknown significance
- Information on fatherhood (primarily as a proof of non-fatherhood)
- Information on incest

A general recommendation is that to a precise, clear question we give a precise answer. The geneticist must, therefore, restrict the report to the question that was posed. There are two exceptions to this rule: when genetic analysis unexpectedly reveals a disease that requires urgent prevention or treatment and in cases of proven incest in underage persons.

The unexpected discovery of a predisposition to a disease represents an additional reason for the requirement that genetic testing always proceeds under the supervision of an experienced physician. Consider the following: a healthy relative of a breast cancer patient has consented to be tested for the BRCA group genes. However, the results of the examination showed a high predisposition to stomach cancer.[3] Should the physician communicate the results of an examination to which the patient has not consented? My personal opinion is that the possibility of unex-

---

[3] See seminar nr. 72.

pected findings should have been presented prior to testing. Without such consent, the physician is left with no good options: either he delivers unwanted burdensome information, or he takes responsibility that the patient will miss the optimal time for early diagnosis and successful treatment.

Genetic examinations in children may provide strong suspicion of incest. A coefficient of homozygosity of approximately 0.5 is convincing evidence that the child's mother was abused by her own father. If possible, the genetic laboratory should repeat the test or (even better) send the data for a confirmatory analysis to another laboratory. In mothers who were underage at the time of conception, we are dealing with proof of a criminal act. The case should, therefore, be reported to the police.

Whenever several family members undergo examinations, genetic testing can show inconsistency with the presumed ("official") fatherhood. In these cases, the results are communicated only if the test for fatherhood was ordered by the court. In case of an accidental finding, I see absolutely no medical, legal, or ethical reasons for communicating discovery of contentious fatherhood to anyone. Revealing such findings would be unfair to women because their infidelity is easier to prove than the infidelity of males. And to take a broader perspective on fatherhood: the father is the one who took care of the child and not someone who encountered his wife at a business dinner party.

Suggested Reading

Anderson JA, Hayeems RZ, Shuman C, Szego MJ, Monfared N, Bowdin S, et al. Predictive genetic testing for adult-onset disorders in minors: a critical analysis of the arguments for and against the 2013 ACMG guidelines. Clin Genet. 2015;87:301–10. https://doi.org/10.1111/cge.12460.

Bird S. Genetic testing: medico-legal issues. Aust Fam Physician. 2014;43:481–2.

Boniolo G, Sanchini V. Ethical counselling and medical decision-making in the era of personalised medicine: a practice-oriented guide. Berlin: Springer; 2016. ISBN-13: 978-3319276885.

Botkin JR, Belmont JW, Berg JS, Berkman BE, Bombard Y, Holm IA, et al. Points to consider: ethical, legal, and psychosocial implications of genetic testing in children and adolescents. Am J Hum Genet. 2015;97:6–21. https://doi.org/10.1016/j.ajhg.2015.05.022.

Chadwick R, Levitt M, Shickle D, editors. The right to know and the right not to know: genetic privacy and responsibility. Cambridge: Cambridge University Press; 2014. ISBN-13: 978-1107076075.

Chiò A, Battistini S, Calvo A, Caponnetto C, Conforti FL, Corbo M, et al. Genetic counselling in ALS: facts, uncertainties and clinical suggestions. J Neurol Neurosurg Psychiatry. 2014;85:478–85. https://doi.org/10.1136/jnnp-2013-305546.

Clarke AJ, Wallgren-Pettersson C. Ethics in genetic counselling. J Community Genet. 2018; https://doi.org/10.1007/s12687-018-0371-7.

DeLisi LE. Ethical issues in the use of genetic testing of patients with schizophrenia and their families. Curr Opin Psychiatry. 2014;27:191–6. https://doi.org/10.1097/YCO.0000000000000060.

Kornyo EA. A guide to bioethics (pocket guides to biomedical sciences). Boca Raton, FL: CRC; 2017. ISBN: 9781138631984.

Dheensa S, Fenwick A, Shkedi-Rafid S, Crawford G, Lucassen A. Health-care professionals' responsibility to patients' relatives in genetic medicine: a systematic review and synthesis of empirical research. Genet Med. 2016;18:290–301. https://doi.org/10.1038/gim.2015.72.

Hall MJ, Forman AD, Pilarski R, Wiesner G, Giri VN. Gene panel testing for inherited cancer risk. J Natl Compr Cancer Netw. 2014;12:1339–46.

Lindor NM, Thibodeau SN, Burke W. Whole-genome sequencing in healthy people. Mayo Clin Proc. 2017;92:159–72. https://doi.org/10.1016/j.mayocp.2016.10.019.

Parker M. Ethical problems and genetics practice. Cambridge: Cambridge University Press; 2012. ISBN-13: 978-1107697799.

Rafiq M, Ianuale C, Ricciardi W, Boccia S. Direct-to-consumer genetic testing: a systematic review of european guidelines, recommendations, and position statements. Genet Test Mol Biomarkers. 2015;19:535–47. https://doi.org/10.1089/gtmb.2015.0051.

Surbone A. Ethical implications of genetic testing for breast cancer susceptibility. Crit Rev. Oncol Hematol. 2001;40:149–57.

Surbone A. Social and ethical implications of BRCA testing. Ann Oncol. 2011;22(Suppl 1):i60–6. https://doi.org/10.1093/annonc/mdq668.

Than NG, Papp Z. Ethical issues in genetic counseling. Best Pract Res Clin Obstet Gynaecol. 2017;43:32–49. https://doi.org/10.1016/j.bpobgyn.2017.01.005.

Vehmas S. Just ignore it? Parents and genetic information. Theor Med Bioeth. 2001;22:473–84.

Abstract

The world of an intensive care unit is a volatile mixture of rapid decisions and actions, great hopes and frequent disappointments, bringing the severely ill back to life and facing death, limited communication with patients, and difficult communication with family members. While some patients recover from severe injuries and apparently hopeless situations, others deteriorate despite all efforts of the medical team. A moment when further intensive treatment is considered to be futile should preferably be made by consensus of the medical team and then communicated to the family. In such a situation, the withholding or withdrawal of intensive medical procedures is justified. Only after confirmation of death are the relatives approached with information on potential organ donation. In the absence of a person's advance directives, some countries allow removal of organs for transplantation regardless of the opinion of family members; nevertheless, a compassionate explanation is preferable and will most often result in consent.

© Springer Nature Switzerland AG 2019
M. Zwitter, *Medical Ethics in Clinical Practice*,
https://doi.org/10.1007/978-3-030-00719-5_16

In 1990, Marina, aged 34, respiratory physiotherapist and mother of five children was diagnosed with acute leukemia. Treatment included intensive chemotherapy and bone marrow transplant, for which a histocompatible donor was found through the Eurotransplant program. Three years later when her cure was confirmed, she expressed a wish to meet her donor. The donor, Terry from Ireland, consented to reveal his identity. A very emotional meeting started a life-long friendship (Photo: courtesy Mrs. Marina A.)

The intensive care unit is a world of its own. An unusual, frequently explosive mix of urgent action, rapid changes, great hopes, successes, and disappointments. A world where being face-to-face with death and coming back to life is common, where communication with patients is limited and communication with patients' relatives often very challenging. The teamwork of intensive care physicians, surgeons, anesthesiologists, and medical nurses is often organized in shifts. This represents an additional obstacle in the transfer of information within the health-care team and with patients and their family and complicates the issue of personal responsibility.

16.1 Admission to the Intensive Care Unit

There is certainly no doubt that a severely wounded or recently operated-on patient belongs in the intensive care unit. The situation becomes more complicated in cases of chronic patients whose health condition has deteriorated to the extent that regular departments can no longer offer safe and professional treatment. Physicians should keep in mind that intensive care units are not intended for dying patients. Whenever colleagues from intensive care units admit a chronic patient with a sudden and potentially treatable deterioration, their intervention should be focused on relieving the acute problem. In cases in which, despite intensive care, the patient's condition does not improve, there must be an option to transfer the patient back to the regular ward in order to continue with palliative care. It is appropriate to inform the relatives about this possibility upon the patient's admission to the intensive care unit. Doing so will make it easier to explain why the patient, despite the worsening of his or her health condition, is no longer treated in the intensive care unit.

Lack of space is the reality for most intensive care units. This reality is taken into account when physicians discuss the admission or transfer of any specific patient. In conversation with the relatives, however, the argument on restricted space in the intensive care unit should not be emphasized as it might give the impression that physicians did not exploit all the possibilities for the patient because they gave preference to other "more important" patients.

16.2 Introducing, Withholding, and Withdrawing Intensive Care

Upon admission to the intensive care unit, the prognosis of a severely diseased patient is often unclear and time for decision-making is limited. Clearly, we will, therefore, do all we can to keep the patient alive.

Only over the following days and weeks will it become clear what course the disease will take. In patients whose condition improved, there are no medical or ethical dilemmas. There are, however, also patients whose health condition deteriorates despite intensive care. This requires careful consideration about whether or not to introduce, withhold or withdraw additional intensive care. The three terms are best illustrated in the following case:

> A patient with advanced ovarian cancer after chemotherapy and radiation therapy has been operated due to intensive vomiting and ileus. She was afterwards admitted to intensive care unit. Her general condition was bad, and blood tests confirmed kidney failure. Physicians could not agree: some argued for the introduction of dialysis while others thought that dialysis would only prolong her suffering and therefore argued for withholding treatment with it. The medical team nevertheless decided in favor of dialysis and to give the patient another chance. After two weeks, however, pleural effusion on lung X-ray examination was interpreted as a clear sign of cancer progression. At that point, the council reached unanimous agreement to withdraw treatment with dialysis. The patient and her relatives were informed about the decision and treatment continued with palliative care.

The difference between withholding and withdrawing treatment is often discussed. From legal and ethical standpoints, there is no difference between these two decisions. Nevertheless, in everyday practice, physicians find the decision not to introduce a specific treatment much easier to make in comparison to a decision to stop a treatment which has already been introduced. This is especially true for treatments that require the use of medical devices, for example, dialysis and mechanical ventilation.

Treatment that must never be withheld or withdrawn is palliative care.

16.3   Futile Intensive Care

When a physician sees that all his skill
And learning can't his patient's death delay
He lets him drink and revel what he may
No more will he prescribe the bitter pill
France Prešeren[1]

Readers have already learned in previous paragraphs what we mean by *futile intensive care*: treatment that has no influence on the basic course of a disease that does not improve the prognosis, and only burdens the patient. Even if such treatment slightly prolonged the patient's life, it nevertheless only prolongs suffering.

I highly recommend that the decision about withholding intensive care be discussed, confirmed, and documented at a meeting of the medical team. This will considerably facilitate conversation with the relatives. We should inform them about our decision compassionately, but we do not ask them for permission or agreement. Nobody, not even the patient himself nor the relatives, can demand the physician to do what he thinks is medically inappropriate, unnecessary, or harmful to the patient—this general principle also pertains to decisions on withholding or withdrawing futile intensive care. At the same time, we should not expect the relatives to confirm or even sign such a decision.

The final remark related to this question: it can happen that the patient's health condition unexpectedly improves. The decision to withhold intensive treatment is not final and can also be revoked.

16.4   Determining Death

In most patients, death is determined without any special diagnostics.

Determining death becomes important in comatose patients sustained by mechanical ventilation and other auxiliary intensive care. If a radioisotope examination confirms the complete loss of blood supply to the brain, there is no doubt: the

---

[1] France Prešeren: Slovenian poet (1800–1849). Translation by Gloria Komai.

patient has died. We do not refer to *brain death*, but *death*. The patient is dead even if the heart will remain active for some time. Even with appropriate levels of compassion, it is not always easy to explain the situation to the relatives; many find it hard to reconcile themselves to the death of their loved one whose appearance is the same as the day before and whose heart is still beating.

Only after death has been confirmed, can we open discussion on possible organ donation.

16.5   Organ Transplantation

To avoid dividing the discussion on transplantation over several chapters, let us first make a detour and briefly describe ethical dilemmas in organ transplantation with living donors. Relevant here are mostly bone marrow and kidney transplantation. In both cases, any form of payment to the donor is unacceptable—organ trafficking is illegal and strongly unethical. The donor performs organ donation out of pure altruism. Most often the donors are adult relatives; underage donors are a rare exception. From an ethical viewpoint, it is important that the proposal for organ donation is not pressured. The potential donor, therefore, has to have the opportunity to give consent during a private conversation with the physician when other family members are not present.

Let us now turn to organ transplantation in the deceased. Once the relatives have been informed about the death of their loved one, we politely inform them about the possibility of organ donation. The intensive care physician should not lead this conversation alone and should introduce a colleague who is in charge of transplantation. We do not want the relatives to have any impression that the physician who treated their loved one has any interest in transplantation; or even worse, that he could hardly wait for their loved one to die.

If the deceased left a written or oral expression of the wish to become an organ donor, the relatives' decision is easier. In most cases, however, such an advance directive is not known. If the relatives have no objections, then we do not demand any signature from them. The process of organ transplantation is carried out through the established protocols. The medical personnel should plan their activities so that the farewell to the deceased does not differ from the usual one in the case of non-donors. When the advance directive of the deceased is not known, and when the relatives oppose organ donation, the process is discontinued. Things become complicated if the patient has clearly expressed an advance directive for organ donation, and the relatives oppose it. We are stranded between the respect for the autonomy of the deceased patient, the potential donor, and the respect for the relatives' wishes. In Slovenia, we follow the wish of the relatives in such cases and, therefore, do not proceed with organ removal. In other countries, a different line of reasoning applies. Consider the practice in Great Britain, where they initially attempt to convince the relatives to respect the advance directive of the deceased. If the negotiation does not succeed, they inform the relatives that they have no legal right to contradict the will of the deceased [1].

Reference

1. https://www.organdonation.nhs.uk/faq/consent/

Suggested Reading

de Groot J, van Hoek M, Hoedemaekers C, Hoitsma A, Smeets W, Vernooij-Dassen M, van Leeuwen E. Decision making on organ donation: the dilemmas of relatives of potential brain dead donors. BMC Med Ethics. 2015;16:64. https://doi.org/10.1186/s12910-015-0057-1.

Godfrey G, Hilton A, Bellomo R. To treat or not to treat: withholding treatment in the ICU. Curr Opin Crit Care. 2013;19:624–9. https://doi.org/10.1097/MCC.0000000000000036.

Goede M, Wheeler M. Advance directives, living wills, and futility in perioperative care. Surg Clin North Am. 2015;95:443–51. https://doi.org/10.1016/j.suc.2014.10.005.

Kostakou E, Rovina N, Kyriakopoulou M, Koulouris NG, Koutsoukou A. Critically ill cancer patient in intensive care unit: issues that arise. J Crit Care. 2014;29:817–22. https://doi.org/10.1016/j.jcrc.2014.04.007.

Lindberg C, Sivberg B, Willman A, Fagerström C. A trajectory towards partnership in care—patient experiences of autonomy in intensive care: a qualitative study. Intensive Crit Care Nurs. 2015;31:294–302. https://doi.org/10.1016/j.iccn.2015.04.003.

Mentzelopoulos SD, Mantzanas M, van Belle G, Nichol G. Evolution of European Union legislation on emergency research. Resuscitation. 2015;91:84–91. https://doi.org/10.1016/j.resuscitation.2015.03.006.

Price S, Haxby E. Managing futility in critically ill patients with cardiac disease. Nat Rev Cardiol. 2013;10:723–31. https://doi.org/10.1038/nrcardio.2013.161.

Quenot JP, Ecarnot F, Meunier-Beillard N, Dargent A, Large A, Andreu P, Rigaud JP. What are the ethical aspects surrounding the collegial decisional process in limiting and withdrawing treatment in intensive care? Ann Transl Med. 2017;5(Suppl 4):S43. https://doi.org/10.21037/atm.2017.04.15.

Rincon F, Lee K. Ethical considerations in consenting critically ill patients for bedside clinical care and research. J Intensive Care Med. 2015;30:141–50. https://doi.org/10.1177/0885066613503279.

Takaschima AK, Sakae TM, Takaschima AK, Takaschima RD, Lima BJ, Benedetti RH. Ethical and legal duty of anesthesiologists regarding Jehovah's Witness patient: care protocol. Braz J Anesthesiol. 2016;66:637–41. https://doi.org/10.1016/j.bjane.2015.03.012.

Thys K, Van Assche K, Nobile H, Siebelink M, Aujoulat I, Schotsmans P, Dobbels F, Borry P. Could minors be living kidney donors? A systematic review of guidelines, position papers and reports. Transpl Int. 2013;26:949–60. https://doi.org/10.1111/tri.12097.

Walton-Moss BJ, Taylor L, Nolan MT. Ethical analysis of living organ donation. Prog Transplant. 2005;15:303–9.

The Elderly and the Mentally-Ill

17

Abstract

Elderly and psychiatric patients share some common characteristics: limited autonomy, uncritical over-estimation of their physical and mental abilities, and need for assistance in daily activities. However, there are also some important differences: psychiatric disease often leads to stigmatisation and to a low social and economic position in society. When dealing with elderly, chronologic age is one important piece of information; yet, age alone should not be a decisive factor in choosing the preferred medical intervention. The main ethical problem is lack of knowledge concerning the elderly: they represent a small minority of patients in clinical trials and are seldom reported separately. In the absence of precise guidelines, we should consider slower metabolic, repair and adaptation processes in advanced age and exert all medical interventions with caution. Caution is also recommended in dealing with psychiatric diseases: every non-conforming behaviour and every minor complaint regarding emotional well-being should not be treated as a disease. Surrogate decision-making should respect the remaining patient's autonomy and should be based on beneficence as the leading ethical principle. When clearly indicated, intervention against a patient's will is ethically justified: effective treatment of the underlying disease will hopefully restore patient's autonomy.

© Springer Nature Switzerland AG 2019
M. Zwitter, *Medical Ethics in Clinical Practice*,
https://doi.org/10.1007/978-3-030-00719-5_17

A thin line, or sometimes no line lies between a genius and a person with psychic disorder. On the picture: self-portrait of Vincent van Gogh (1853–1890), a Dutch painter who suffered from severe psychotic episodes

We will treat ethical questions pertaining to the elderly and the mentally-ill under the same heading because their issues have much in common. In both groups, a partial or complete loss of autonomy can occur, leading to surrogate decision-making or guardianship. In both groups, there are individuals in need of assistance. In addition, both the elderly and the mentally-ill often lack critical judgment and overestimate their competence. In senile dementia, both conditions merge.

We must not overlook, however, significant differences between the two groups. Mental disease affects only certain members of society, but aging happens to everyone. The two groups also differ markedly in their social status. We treat the elderly with respect, and we like to offer them help, but we often remain reserved towards the mentally ill and prefer to keep them out of our social circles. The economic status is different as well. The elderly will often accumulate some wealth in addition to receiving a decent pension. The mentally ill, however, often find themselves at the margins of society and struggle to make ends meet with a low disability pension.

17.1   Who Are the Elderly?

There is no sharp, well-defined boundary between adults and the elderly. We also do not want to define one here. Whatever boundary point we choose—be it 65, 75 or 85 years—we will have no trouble finding exceptions in both upwards and downwards directions. Without considering other factors, the chronological age alone, therefore, cannot be the criterion in the physician's decision-making. However, it would be unreasonable to completely abandon age as *one of the factors* influencing the risk of developing a disease, the natural course of the disease, and the success or complications of a treatment.

In the eyes of a physician, an elderly person belongs to one of two groups. In the first group are those who had a hard life, which has affected their current health condition. A number of chronic diseases, often accompanied by a decline in cognitive functions and in vital energy all gradually lead to an elderly person lacking the capacity for autonomous decision-making and greatly dependent on assistance. The second group comprises the elderly who can be likened to a well-maintained classic car: looks great, works great! Think of the witty and energetic interviews the Slovenian writer Boris Pahor gave when he was 103 years old!

Regardless, my advice is to be careful when treating an elderly person from this group. Blood examinations or an electrocardiogram do not reveal everything. The analogy with the classic car was not chosen at random: a defect in an old car will often start a cascade of new defects. It is often similar with the elderly: aggressive treatment that might be well-suited for younger patients can often turn out to be problematic in the elderly. One can often read in professional literature about how certain medications have no age limits and are suitable for elderly patients with normal indicators for specific organ systems (blood tests, kidney and liver function, cardiovascular systems). However, we must pay attention to three specific circumstances. First, clinical trials often include limited numbers of elderly patients, so we often simply do not have the data to judge the safety of many medications (recall the discussion of a similar blind spot in pediatrics). Second, an important practical consideration lies in realizing that the pharmaceutical industry has a strong interest in expanding their markets to the ever-growing population of the elderly. Third, the metabolism for many medications slows down considerably with age. A prudent physician will therefore introduce new treatment carefully and gradually.

Let us briefly return to the topic of chronological age as a criterion for the physician's decision-making. In principle, it is, of course, wrong to set up a harsh age limit for certain very expensive or complicated treatments or for assigning priority in treatment. Therefore, we do not categorize the patients according to their age; we rely on the diagnosis, severity of the disease, and urgency instead. In medical practice, however, things change. In a medical team running short of all resources—time for patients, medical devices, financial resources for medications, hospital beds—age becomes one of the factors in decision-making. Consider a priority list for heart transplantation: since the number of potential recipients far exceeds the number of organs available for transplantation, the age of the recipient is one of the factors that determine his/her position on the list. For health inspectors reading this—yes, this is how things work: physicians simply should not loose their humanity and empathy in cold grading systems. We all love our elderly patients just the same, however, I would have trouble understanding a physician who would not react emotionally when facing a severe disease in a young mother.

Whatever age criterion we agree upon, the fact remains that the population is aging. How should the healthcare system react to this issue? A detailed answer is obviously beyond the scope of this chapter and book; however, we can simplify the issue somewhat and give the following answer: we do not need more beds in intensive care units; instead, we need significant improvements in home care and in a network of hospices.

17.2   Who Are the Mentally Ill?

For some of the mentally ill, the presence of the disease is beyond doubt. Because this book is not a psychiatry textbook, I shall not delve any deeper in diagnostic procedures and diseases. At this point, we will have to content ourselves with the realization that a significant number of mentally ill patients need assistance with everyday chores as well as with treatment-related decision-making.

Still, any sort of deviation does not yet mark a disease. One must acknowledge the fact that some people see their sojourn on the planet Earth differently from the way the majority does. Only a few decades ago, we deemed homosexuality or political dissent to be mental illnesses, and in some parts of the world this remains the case. Is someone who feels no attachment to wealth or regular employment mentally-ill? Should numerous artists be treated as mentally-ill? Are moments of disappointment, sadness, unease, sleeplessness, despair or fierce anger themselves mental illnesses? Or are these moments common to all of us?

We must therefore be cautious when making a diagnosis of a mental illness. The physician should take special care in judging whether the patient really needs treatment with medications or not. Yet again, I have to point out the interests of the pharmaceutical industry, which encourages the "invention" of novel types of diseases and tries to sell drugs for them. Addiction to anxiolytics, tranquilizers, and sleeping pills is very common, especially among healthcare workers and pharmacists.[1] Interesting measures have been adopted in New Zealand, where a physician can, instead of medication, prescribe a book for a troubled or distressed patient (costs covered by health insurance).[2]

We must also mention a special category of mentally-ill patients: alcoholics and illicit drug addicts. Given suitable help and support, some can overcome the burden of addiction for good. The first big and meaningful step is to admit one's own addiction, for which a positive reaction from the environment is of crucial importance. Among the highly-educated, even among physicians, it is not uncommon to observe false comradeship and concealment of addiction, which of course blocks all possible paths to a solution.

---

[1] See seminar nr. 43.

[2] See seminar nr. 44.

17.3 Impaired Critical Judgement and Loss of Autonomy

A common point between the elderly and the mentally-ill is overestimating, or sometimes under-estimating one's own competence. Excessive confidence can have negative consequences for the elderly or mentally-ill patient him or herself, but clearly also for his or her close and remote social environment. Think of the issue of driving: a driver lacking driving skills or a drunk driver represents a threat to himself and also to all those nearby.

Situations like these put close friends and relatives in a dilemma. They do not want to hurt the person's feelings, but at the same time they are concerned about the safety of everyone involved, for example their children.

Concealing the problem or, as we often say, sweeping the issue under the rug, is not only a problem with the individual; unfortunately, it also happens at the level of society as a whole, suggesting the apparent unwillingness of society to bear its share of responsibility. The legislation often shows exaggerated fears of potentially doing injustice to the individual, but fails to protect other citizens. In many countries, including my own, a driving license is valid until the age of 80. Every year, a few elderly drivers are victims of traffic accidents due to driving on a highway in the wrong direction. From the medical standpoint, we know that few octogenarians are indeed able to drive safely, and I am therefore strongly in favor of mandatory testing for the ability to drive safely for all persons above a certain age (e.g. 65 years). Likewise, chronic alcoholics or those who exert their aggressive personality through high-risk driving should be strictly controlled, and permanent revocation of the driving license should be a regular practice, rather than a rare exception. A decision about the ability to drive safely should fall upon the authorities, rather than upon the person's friends, relatives, or the family physician.

17.4 Surrogate Decision-Making

When a patient is not able to completely understand his or her own situation and make decisions, the physician will seek consensus with the patient's close friends and relatives. In most cases, guardianship is not (yet) officially confirmed. If important decisions are to be made, the family should agree about who is to make decisions in the patient's name.

We have already discussed surrogate decision-making in the chapter on autonomy, so let us just briefly recap here: surrogate decision-making depends on the ethical principle of beneficence, rather than the principle of autonomy. The physician, therefore, is not bound to respect the decisions of the patient's relatives if these are clearly against the patient's best interest. However, in such a case, the physician should not decide alone and is strongly advised to seek support for his or her decisions from other colleagues.

When an agreement with the relatives is reached and decisions are made, the physician still has the duty to inform the patient. All further procedures should be explained in a way that is clear and understandable for the patient. As long as the patient is conscious, the loss of autonomy is not complete, and we should therefore encourage his or her understanding of the situation and attempt to obtain his or her opinion as well. The patient should make his own decisions regarding the domains he can still understand.

Relations with relatives are often strained or explicitly conflictual due to their demands which, though in the patient's interest, prove to be unrealistic. An example of such a conflict is the restraints that are used to prevent the restless and unmanageable patients falling out of their beds. Relatives oppose these measures and instead expect a nurse to be present at the patient's bed at all times—an unrealistic demand if, during a night shift, one or two nurses are responsible for an entire hospital ward. It is the duty of the physician to defend the nurse and explain the pragmatic facts of the circumstances to the relatives.

17.5   Acting Against the Will of the Patient

An elderly, extremely feeble patient lived alone in a remote house with no central heating. The closest neighbor was 1 km away, and even farther were the nearest store, pharmacy and his community physician. The patient steadfastly refused the proposal to move to a retirement home after discharge from the hospital. Instead, he insisted he would go home after his hospital stay.

This true story is a good example of a patient overestimating his own competence—a topic that we have touched upon already—and sets a good starting point for discussion of situations requiring actions against the will of the patient. The elderly often show strong prejudice against retirement homes. They do not understand that for them these housing facilities offer significantly higher standards of living, safety and social contacts. In fact, loneliness is frequently a huge burden for the elderly, especially for widows. Despite the patient's opposition, referral to a retirement home is often the only possible solution. In a retirement home, the elderly patient will make new friends and the traumatic experience of moving out of home will soon be forgotten.

The rules regulating the involuntary hospitalization of psychiatric patients are clearly stated in the legislation. From the perspective of ethics, two points should be stressed. The first refers to psychiatric diagnosis. As emphasized at the outset of this chapter, any deviation from normality in itself does not yet constitute a mental illness and as such is definitely not a sufficient reason for involuntary hospitalization. Second, when mental illness is clearly diagnosed in a bewildered and delusional patient, it must be pointed out that in this case we are not talking about acting against the patient's autonomy. It is the disease itself that is the cause of the patient's lack of autonomy, and medical treatment will restore the person's autonomy.

17.6   Hunger Striking and Anorexia Nervosa

My apologies for including the section on hunger striking and anorexia nervosa in this chapter. For most people, a hunger strike is a very unusual form of protest, yet it cannot be considered a sign of a mental disease. Anorexia nervosa is listed among mental diseases; yet, it should not be automatically categorized to the group of diseases that deprive an individual of his autonomy, with the consequent right to act against his will. In both cases, the physician deals with an autonomous person who should be allowed to decide freely regarding his or her life.

During the initial period of a hunger strike, when life is not yet at stake, the role of a physician should be limited to advice and counselling. Infusions and force-feeding are acceptable only in cases of real and immediate danger to life. In anorexia nervosa, the situation is somewhat different. Even before being in immediate danger of irreversible deterioration of their health, some patients recognize their condition as a disease and accept treatment and feeding. Yet, even here, any intervention against the patient's will is justified only in case of danger to life.[3]

17.7 Social Status and Stigmatization

Thus far, we have considered similarities between the elderly and the mentally-ill patients. Here we consider points of considerable divergence between the two groups. While the elderly are treated respectfully and with affection, this is rarely the case for the mentally ill. Many of the elderly are financially secure, whereas this is rarely the case with the mentally ill.

Despite great progress in the treatment of mental illnesses, these patients are still often stigmatized. We could have a long unfruitful discussion about how wrong this is. It is more productive to focus our attention to the question of what can be done to limit stigmatization and make it less painful. When compared with the old practice of long-lasting hospitalization, out-patient treatment of psychiatric patients is clearly a positive approach. Yet, not much has been done if the medical assistance is limited to office consultations and the prescription of drugs. Apart from physicians, social workers, psychologists, and nurses should aim towards comprehensive rehabilitation and greater integration of patients into society. This would lead to a lower proportion of patients with relapses of their mental disease and gradually to the reduced stigmatization of these patients in society.

Suggested Reading

Allen NG, Khan JS, Alzahri MS, Stolar AG. Ethical issues in emergency psychiatry. Emerg Med Clin North Am. 2015;33:863–74. https://doi.org/10.1016/j.emc.2015.07.012.

Daher M. Ethical issues in the geriatric patient with advanced cancer 'living to the end'. Ann Oncol. 2013;24(Suppl 7):vii55–8. https://doi.org/10.1093/annonc/mdt262.

Diekema DS. Involuntary sterilization of persons with mental retardation: an ethical analysis. Ment Retard Dev Disabil Res Rev. 2003;9:21–6.

Gauthier S, Leuzy A, Racine E, Rosa-Neto P. Diagnosis and management of Alzheimer's disease: past, present and future ethical issues. Prog Neurobiol. 2013;110:102–13. https://doi.org/10.1016/j.pneurobio.2013.01.003.

---

[3] See seminar Nr 14.

Gopal AA. Physician-assisted suicide: considering the evidence, existential distress, and an emerging role for psychiatry. J Am Acad Psychiatry Law. 2015;43:183–90.

Hébert PC, Weingarten MA. The ethics of forced feeding in anorexia nervosa. CMAJ. 1991;144:141–4.

Heinimaa M, Larsen TK. Psychosis: conceptual and ethical aspects of early diagnosis and intervention. Curr Opin Psychiatry. 2002;15:533–41.

Ho AO. Suicide: rationality and responsibility for life. Can J Psychiatr. 2014;59:141–7.

Hughes J, Common J. Ethical issues in caring for patients with dementia. Nurs Stand. 2015;29:42–7. https://doi.org/10.7748/ns.29.49.42.e9206.

Humphreys K, Blodgett JC, Roberts LW. The exclusion of people with psychiatric disorders from medical research. J Psychiatr Res. 2015;70:28–32. https://doi.org/10.1016/j.jpsychires.2015.08.005.

Irmak N. Professional ethics in extreme circumstances: responsibilities of attending physicians and healthcare providers in hunger strikes. Theor Med Bioeth. 2015;36:249–63. https://doi.org/10.1007/s11017-015-9333-9.

Kukreja D, Günther U, Popp J. Delirium in the elderly: current problems with increasing geriatric age. Indian J Med Res. 2015;142:655–62. https://doi.org/10.4103/0971-5916.174546.

Lazarus JA. Physicians' ethical obligations to hunger strikers. BMJ. 2013;346:f3705. https://doi.org/10.1136/bmj.f3705.

Silber TJ. Treatment of anorexia nervosa against the patient's will: ethical considerations. Adolesc Med State Art Rev. 2011;22:283–8.

Strand M, von Hausswolff-Juhlin Y. Patient-controlled hospital admission in psychiatry: a systematic review. Nord J Psychiatry. 2015;69:574–86. https://doi.org/10.3109/08039488.2015.1025835.

Svensson B, Hansson L. How mental health literacy and experience of mental illness relate to stigmatizing attitudes and social distance towards people with depression or psychosis: a cross-sectional study. Nord J Psychiatry. 2016;70:309–13. https://doi.org/10.3109/08039488.2015.1109140.

Dying and Death

18

Abstract

A hundred years ago, sitting alongside a dying family member was a normal experience. Nowadays, death is considered to be an undesirable guest in our homes, and most people die in hospitals or hospices. While the sentence "I wish I would die" is often heard, it should be interpreted as a call for help, rather than as a demand for euthanasia. Surveys in countries where euthanasia or physician-assisted suicide are legal revealed that the demand of the dying person is often a consequence of social, emotional, or even financial circumstances. Euthanasia should never become a shortcut to relieve society of the burden of people approaching the end of their life. In circumstances of severe deficiencies in palliative care, a debate on euthanasia is ethically problematic. Investment in facilities for palliative medicine and in the related activities in education and research should become a priority. Only when the goal of decent palliative care for everyone is achieved, will a discussion on euthanasia truly become a discussion on individual autonomy, rather than a discussion based on utilitarian ethics.

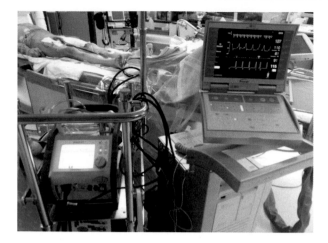

Towards death with dignity. Ethically problematic are both extremes—on one side active euthanasia, and on the other treatment of terminally ill patients with all procedures of intensive medicine which just prolong suffering (Photo: Dr. Marko Noč)

Death must never be taken lightly, even if one has seen it hundreds of times. I am considerably less bothered by an admission of professional malpractice (there is none among us who has not made a mistake) than by a casual or even derisive attitude towards death. It hurts deeply to hear someone starting their report on overnight duty by saying they had "a bumper crop" or if I hear nurses saying that some physicians do not allow nurses to wake them if a patient dies in the middle of the night. Here is a real example of a colleague of mine, who also departed his life long ago and who did not allow himself to be disturbed when he was on night duty. It happened that two patients passed away during the night and were taken to the mortuary. The next morning, he was signing the documentation, after which a nurse incorrectly attached the identification labels. As a consequence, the deceased from Bosnia was almost buried in the Slovenian town of Piran. This is not only a severe violation of work responsibilities, but also disrespectful of the deceased, of other patients in the same department, and of the nurses.

In recent years, death has been discussed more frequently in public. Although the topic of dying is no longer a taboo, many still have great difficulty accepting the passing of their loved ones. Regardless of how clear it is that the patient is terminally ill, the patient's relatives will still all too often bring the patient to the emergency department. Simple people who keep to their rural family traditions (we were all farmers at some point) still gather the courage and spirit to sit with their dying father at home; people from higher classes of society, however, start panicking about how death will spoil their perfect world.

We should be comfortable with the fact that our death and that of others is inevitable. All religions include a promise of the afterlife. There is no doubt that this helps many to leave this world in peace. Our duty is, therefore, to offer spiritual support to the severely ill. To those among us who are not endowed with such a deep and concrete religiosity, serenity comes from an awareness that we are not leaving behind a world of injustice or unpaid dues and that we have at least strived towards a better world.

18.1   The Wish to Die and Suicide

"Doctor, I would rather just die" is a sentence we physicians hear frequently. Clearly, we should not take these words literally but as a call for help: pain, other physical troubles, fatigue, the inability to take care of oneself, and loneliness. In patients with diseases such as cancer, a genuine wish to die is rare, at least in my experience. Even severely ill patients wish to live, and the more life is moving away from them, the more they cling to it. After all, this is reflected in statistics on suicide in oncology patients, which may be even less frequent than in the healthy population.[1]

The phenomena of attempted suicide and suicide can only be discussed briefly in this book, even though they would require a much more in-depth treatment. In the discussion of ethical questions, we must first bring up the responsibility for poor social conditions that lead an individual to suicidal thoughts and to commit the act. It would be too non-committal to claim that we are all guilty of these conditions. Let us consider a recent suicide of an entrepreneur who did not receive payment for his work and had to declare bankruptcy, which, in turn, caused severe personal distress. The responsibility for the suicide can be rightly attributed to the ministers for finance, economy, and justice—and even all the way up to the prime minister—who allow and sometimes even encourage financial disorder. The leading politicians are to be blamed for the severe stratification of society: on the one side, we are witnessing the illegal accumulation of wealth while on the other side, poverty, social distress, and alcoholism bring about despair, sending some individuals over the edge.

What is the physician's role in suicide prevention? As in all branches of medicine, the physician should raise awareness about the factors in disease and premature death. Those physicians who left medicine to engage in national politics or administration have a special responsibility because they are actively directing our societal activities. In cases of suicide attempts, the family physician or the psychiatrist will contact the social welfare services and will attempt to relieve the personal distress of the unfortunate person. Finally, we must not forget the loved ones: family, friends, and co-workers. People in distress frequently show signs of suicidal tendencies well in advance, but we either do not hear their calls for help, we do not take them seriously, or we are not able to take action. After suicide, the entire circle of the deceased experiences a shock and the feeling of guilt. It is true that any death echoes in pain; however, relatives' distress after suicide is especially severe.

---

[1] I have the data for the Institute of Oncology in Ljubljana for years 1995–1999 when I was director: one suicide and one suicide attempt. For 320 hospital beds and close to 100,000 total days of hospitalization per year, this is not a big number, even if we take into account unrecognized suicides, for example due to overdoses.

18.2   What Is Not Euthanasia?

The term *passive euthanasia* is inappropriate and misleading; therefore, my advice is to not use it. As we explained in the chapter on intensive medicine, withdrawing futile intensive treatment in terminally ill patients is medically and ethically the only correct measure and should therefore not be considered a kind of euthanasia. The majority of oncology patients could have had their lives prolonged for a short time, for example, by using artificial respiration, dialysis, intensive supportive care (infusion, transfusion, antibiotics in terminal pneumonia, correction of electrolyte imbalance, nasogastric intubation for nutritional support); however, we would only prolong their suffering. In the words of our teacher Janez Milčinski[2]: "If you are playing chess with a good chess player and you lose the queen, you will surrender the match. And believe me, death is a very good chess player." A compassionate physician will put his efforts into relieving the physical and mental problems of the patient and offering spiritual support, rather than to unreasonable and painful prolongation of life. Clearly, such a decision has nothing to do with euthanasia.

In the past, we also talked about treatment with dual effects. While certain medications such as morphine can alleviate pain, high doses lead to depression of breathing and hasten the death of the patient. Today, we know that morphine is a relatively safe drug and is one of the drugs of choice in pulmonary edema, the condition in which respiratory distress is the leading symptom. Relieving patient's pain will make him less upset and will, therefore, alleviate respiratory distress. With appropriate dosage, there is thus no danger of causing respiratory depression with morphine.

The term *euthanasia without patient's consent* is also inappropriate. In this case, we can have a competent adult person who could have required euthanasia but has not done so (some refer to this as *involuntary active euthanasia*) or a person who cannot make such a demand, for example, a child, an unconscious patient, or a patient with a mental disorder (this is supposedly termed *nonvoluntary active euthanasia*). Such wordsmithery does not change the fact that these are all instances of murder.

18.3   Euthanasia and Physician-Assisted Suicide

Euthanasia is a procedure in which a person (normally a physician) administers to another competent adult person, at that person's explicit demand, a drug (e.g., a sedative or a muscle relaxant) with the purpose of ending the person's life. In *physician-assisted suicide*, the physician prescribes or provides a drug to another competent person, at that person's explicit demand, and is aware that the person will use the drug to end his or her own life.

Euthanasia has been legalized in the Netherlands, Belgium, Columbia, and Canada. Physician-assisted suicide, which does not include euthanasia, is allowed

---

[2] Janez Milčinski: physician and legal expert, one of the fathers of medical ethics in Slovenia.

in Switzerland and in some states of the USA [1]. Before we delve into a discussion of the ethical aspects, let us consider the experience of some of the countries where euthanasia is allowed.

In the Netherlands, euthanasia was used to end the lives of more than 5500 people in 2015, which amounts to approximately 4% of all deaths. Roughly 70% of the cases involved advanced cancer patients. Euthanasia was also performed on a 20-year-old woman who suffered from severe psychological issues due to sexual abuse in childhood and in a 41-year-old man who was not able to overcome his chronic alcoholism [2, 3]. Mental disorders were specified as an indication in 6% of all euthanasia cases in the Netherlands. The number of persons on whom euthanasia is legally performed due to mental disorders is rising steeply.

It is also interesting to see the motivation for demanding euthanasia or physician-assisted suicide. Inadequate pain control was provided as the direct cause in 49% of the patients in the Netherlands. In American patients, among the end-of-life concerns for physician-assisted suicide were the loss of autonomy and joy of life (90%), loss of dignity (76%), the burden for family or friends (53%), pain (36%), or financial troubles (9%).[3]

All the above-mentioned experiences are clearly indicating the broad field of abuse that opens up with the legalization of euthanasia and physician-assisted suicide. The right to autonomy, often pointed out by the advocates of euthanasia, cannot apply to persons who do not have it, for example, individuals suffering from mental disorders or children. It is also highly debatable whether a patient suffering from symptoms of advanced cancer is fully autonomous.

Proponents of the legalization of euthanasia and of physician-assisted suicide emphasize the ethical principle of autonomy and claim that the right of an individual to decide about his or her life and death is a basic human right. As we have already said, it is debatable whether a person suffering from a terminal disease is genuinely autonomous. In rare instances of mentally fully competent persons who reached a decision to end their life, the legalization of euthanasia and of physician-assisted suicide may indeed be regarded as a positive shift towards the full respect of their autonomy. However, we should be aware of the inevitable abuse, both in the person's micro-environment and in the nation at large. Experience from countries where the procedures are legal shows that motivation for demanding euthanasia or physician assisted suicide is often societal, rather than personal. While we do not have exact data for the Netherlands, approximately half of the USA patients submit a request for physician-assisted suicide because they believe that burden their loved ones, be it in terms of care or even costs. In all these cases, patients do not enjoy full autonomy. Rather, their decisions are put under pressure due to inadequate treatment of their symptoms or due to problems in care.

Absolutely unacceptable are all forms of involuntary euthanasia—we have already stated that these are instances of murder.

Abuse on a larger, national scene should also be considered: legalization of euthanasia instead of offering free facilities for palliative care. By all means,

---

[3] More than one motivation factor for demanding physician-assisted suicide may apply.

euthanasia must not serve as a convenient and cheap way of alleviating people's distress because the availability of palliative care is limited. This conclusion certainly does not apply only to the USA or the Netherlands, but to many other countries as well: it is highly inappropriate to open the discussion on euthanasia in countries where the development of palliative medicine is slow and ineffective. In my opinion, any moves to legalize euthanasia will be acceptable once we ensure that everyone is offered free palliative care at home or in a nursing home. Personally, I will oppose euthanasia in such cases as well. I will, however, concede to the proponents of euthanasia that it is, in fact, a question of the individual's autonomy rather than the society's attempt to take the shortcut in dealing with the burden of the terminally ill.

Performing euthanasia or physician-assisted suicide is in clear opposition to the mission of the physician, and it goes against all written ethical documents, from Hippocrates onwards. If euthanasia is legalized—an act that physicians cannot prevent—I am confident that many physicians will resort to conscientious objection. It is true, however, that one does not need a medical education to kill a person. Parliament should therefore also specify who should be performing euthanasia. There are quite some philosophers among the proponents of euthanasia; this is their job opportunity.

18.4   Persistent Vegetative State

Despite my disapproval of euthanasia, I cannot ignore an exceptional case in which euthanasia might be acceptable under certain conditions: persistent vegetative state following brain injury. Given proper care, feeding through a stomach tube, and the prevention of bedsores, these patients can remain in a stable condition for several years; however, they remain so without noticeable signs of self-awareness and with no means of communication. Not only is sustaining life at such a low level a burden for the nursing staff and for those who support it financially, but it also represents a significant psychological burden to the relatives.

> After an injury in a traffic accident, Eluana Englaro remained in a vegetative state for 17 years. At the request of her parents and in line with the court decision, feeding and supply of liquids were discontinued. After one week, on February 9, 2009, she passed away.

Before and after Eluana's death, a heated discussion developed about the issue of whether feeding and supply of liquids should be viewed as medical measures that can be withdrawn in a patient with no hope of recovery, or whether the two measures should instead be considered as part of the basic care that must be guaranteed to anyone regardless of his or her health condition and prognosis. This discussion, however, missed the point: this was an example of euthanasia, albeit with a time delay and unnecessary suffering while the patient was dying from hunger and thirst. They unplugged the feeding and hydration tubes and pretended they were not performing euthanasia; it was as if they said: "Eluana, the fridge is over there. If you are hungry or thirsty, please help yourself."

A similar condition is irreversible brain coma after brain stroke. It sometimes happens that physicians will unplug a hopelessly ill unconscious patient from the apparatus for artificial respiration and then let him die over the next hours while gasping for air. If the physicians in both presented cases decided that the patient's health condition is hopeless and if they decided to discontinue feeding or breathing support, then they should also take the next step and ensure the immediate death of such a patient without additional suffering.

However, we cannot ignore certain restrictions with respect to whether or not euthanasia is acceptable in a persistent vegetative state or for comatose patients. First, unequivocal confirmation of the hopeless condition must be established. This should be entrusted to a medical team and not to an individual physician. Euthanasia must not be performed without consent or even against the will of the relatives. Here, the relatives assume the role of surrogate decision-makers. Without their explicit agreement, one would perform euthanasia without consent (*nonvoluntary active euthanasia*), which we designate as a murder. Finally, there is the question of autonomy and conscientious objection of the physician, who at all times retains the right to decline euthanasia.

18.5 On Immortality

The opposite of the wish to die, that is, suicide or euthanasia, is the wish for immortality.

Numerous observations in humans and animals suggest that the process of aging and the length of life are part of our genetic code. Such biological limits make sense—the elderly must make way for the younger generation in order to ensure the progress and survival of the species.

Prolongation of life, as seen through statistics, does not in itself prove that we will be able to live up to 130 or 150 years in the future. These statistics describe trends in average life expectancy. Due to improved living conditions in most parts of the world, average life expectancy is indeed increasing. At the same time, it is true that as early as in the Middle Ages, noblemen and other wealthy individuals lived quite long lives, even over 70 years—provided, of course, they did not contract any of the diseases that were fatal in those days. What we have seen in the last century is an increase in the average life expectancy, rather than the prolongation of the longest survival.

A physician offering advice on healthy life in old age is supporting the desire of people to live longer and especially better. Undoubtedly useful are certain fairly simple measures that can help to resolve acute or chronic issues and significantly improve the quality of life in the elderly. I do recommend, however, a great deal of skepticism when it comes to efforts to decelerate aging with medications or various nutritional supplements: there is no evidence of their efficacy, and the only ones who definitely benefit from these measures are the manufacturers and vendors of these wonder pills and drinks. Indeed, fear of death is a fertile ground for profiteers of all kinds. In the confused world of hunger and poverty on one side, and of extreme wealth on the other, some can afford being frozen upon dying and thus hope for a new life once the progress in medicine will allow successful treatment of their disease. How perverse to supplant serene and peaceful farewell with consumerism, which is being infiltrated into such an intimate period as dying.

18.6   Education and Research

For obvious reasons, dying and death are not very interesting topics in classrooms and in research. In oncology as well as in other fields of medicine, textbooks and everyday practice offer the same picture: some words on preventive measures, much attention to basic research, diagnostics and specific treatment, few words about palliative care, and silence about the final period of life. While all physicians can easily recall the individual patients in their final days of life, comprehensive data and serious analyses are scarce. Without a reliable scientific analysis, problems are not identified, and a plan to resolve the pressing needs cannot be presented.

After our discussion in this chapter, it is clear that physicians should devote more attention to the final period in life, dying, and death. Even though many of us reject euthanasia as a medical procedure for immediately causing death, we at the same time endorse euthanasia in its literal translation from Greek: euthanasia is the right to a decent and good death.

References

1. Emanuel EJ, Onwuteaka-Philipsen BD, Urwin JW, Cohen J. Attitudes and practices of euthanasia and physician-assisted suicide in the United States, Canada, and Europe. JAMA. 2016;316:79–90. https://doi.org/10.1001/jama.2016.8499.
2. http://www.cbsnews.com/news/netherlands-sex-abuse-victim-euthanasia-incurable-ptsd-assisted-suicide/
3. http://www.dailymail.co.uk/news/article-3980608/Dutch-euthanasia-law-used-kill-alcoholic-41-decided-death-way-escape-problems.html

Suggested Reading

Baider L, Surbone A. Patients' choices of the place of their death: a complex, culturally and socially charged issue. Onkologie. 2007;30:94–5.
Barutta J, Vollmann J. Physician-assisted death with limited access to palliative care. J Med Ethics. 2015;41:652–4. https://doi.org/10.1136/medethics-2013-101953.
Boudreau JD, Somerville MA. Euthanasia is not medical treatment. Br Med Bull. 2013;106:45–66. https://doi.org/10.1093/bmb/ldt010.
Branigan M. Desire for hastened death: exploring the emotions and the ethics. Curr Opin Support Palliat Care. 2015;9:64–71. https://doi.org/10.1097/SPC.0000000000000109.
Chambaere K, Bernheim JL. Does legal physician-assisted dying impede development of palliative care? The Belgian and Benelux experience. J Med Ethics. 2015;41:657–60. https://doi.org/10.1136/medethics-2014-102116.

Cholbi M. Kant on euthanasia and the duty to die: clearing the air. J Med Ethics. 2015;41:607–10. https://doi.org/10.1136/medethics-2013-101781.

Cohen-Almagor R. First do no harm: intentionally shortening lives of patients without their explicit request in Belgium. J Med Ethics. 2015;41:625–9. https://doi.org/10.1136/medethics-2014-102387.

Connolly S, Galvin M, Hardiman O. End-of-life management in patients with amyotrophic lateral sclerosis. Lancet Neurol. 2015;14:435–42. https://doi.org/10.1016/S1474-4422(14)70221-2.

Dewar R, Cahners N, Mitchell C, Forrow L. Hinduism and death with dignity: historic and contemporary case examples. J Clin Ethics. 2015;26:40–7.

Gillon R. Sanctity of life law has gone too far. BMJ. 2012;345:e4637. https://doi.org/10.1136/bmj.e4637.

Gillon R. Why I wrote my advance decision to refuse life-prolonging treatment: and why the law on sanctity of life remains problematic. J Med Ethics. 2016;42:376–82. https://doi.org/10.1136/medethics-2016-103538.

Jones J. Do not resuscitate: reflections on an ethical dilemma. Nurs Stand. 2007;21:35–9.

Landry JT, Foreman T, Kekewich M. Ethical considerations in the regulation of euthanasia and physician-assisted death in Canada. Health Policy. 2015;119:1490–8. https://doi.org/10.1016/j.healthpol.2015.10.002.

Luce JM. Chronic disorders of consciousness following coma: part two: ethical, legal, and social issues. Chest. 2013;144:1388–93. https://doi.org/10.1378/chest.13-0428.

Mallia P, Daniele R, Sacco S, Carolei A, Pistoia F. Ethical aspects of vegetative and minimally conscious states. Curr Pharm Des. 2014;20:4299–304.

Marx G, Owusu Boakye S, Jung A, Nauck F. Trust and autonomy in end of life: considering the interrelation between patients and their relatives. Curr Opin Support Palliat Care. 2014;8:394–8.

Mesquita AC, Chaves ÉCL, Barros GAM. Spiritual needs of patients with cancer in palliative care: an integrative review. Curr Opin Support Palliat Care. 2017;11:334–40. https://doi.org/10.1097/SPC.0000000000000308.

Pérez Mdel V, Macchi MJ, Agranatti AF. Advance directives in the context of end-of-life palliative care. Curr Opin Support Palliat Care. 2013;7:406–10. https://doi.org/10.1097/SPC.0000000000000007.

Rabiu AR, Sugand K. Has the sanctity of life law 'gone too far'?: analysis of the sanctity of life doctrine and English case law shows that the sanctity of life law has not 'gone too far'. Philos Ethics Humanit Med. 2014;9:5. https://doi.org/10.1186/1747-5341-9-5.

Ramondetta LM, Sun C, Surbone A, Olver I, Ripamonti C, Konishi T, et al. Surprising results regarding MASCC members' beliefs about spiritual care. Support Care Cancer. 2013;21:2991–8. https://doi.org/10.1007/s00520-013-1863-y.

Rodríguez-Prat A, Monforte-Royo C, Porta-Sales J, Escribano X, Balaguer A. Patient perspectives of dignity, autonomy and control at the end of life: systematic review and meta-ethnography. PLoS One. 2016;11:e0151435. https://doi.org/10.1371/journal.pone.0151435.

Rys S, Deschepper R, Mortier F, Deliens L, Atkinson D, Bilsen J. The moral difference or equivalence between continuous sedation until death and physician assisted death: word games or war games?: a qualitative content analysis of opinion pieces in the indexed medical and nursing literature. J Bioeth Inq. 2012;9:171–83. https://doi.org/10.1007/s11673-012-9369-8.

Walker P, Lovat T. Life and death decisions in the clinical setting: moral decision making through dialogic consensus. Berlin: Springer; 2017. ISBN-13: 978-9811043000.

Walton L, Bell D. The ethics of hastening death during terminal weaning. Curr Opin Crit Care. 2013;19:636–41. https://doi.org/10.1097/MCC.0000000000000027.

Weinberger LE, Sreenivasan S, Garrick T. End-of-life mental health assessments for older aged, medically ill persons with expressed desire to die. J Am Acad Psychiatry Law. 2014;42:350–61.

Research

19

Abstract

To relieve human suffering—the motto of our profession—extends beyond our current patients to those in the future. The progress of medicine depends on medical research. Our discussion will be limited to clinical trials with critical ethical issues due to the active involvement of patients. From the patients' standpoint, the essential demands are: a scientifically sound research plan; free choice to enter a trial or receive routine treatment; understandable informed consent; and, in the case of randomized trials, physician's equipoise (sincere uncertainty) regarding preferences to the proposed treatments. Despite all efforts, some ethical costs are inevitable: in comparison to routine treatment, clinical research often includes additional diagnostics, the collection of personal data, and limited freedom to individualize treatment. Publication of results serves to counterbalance these inevitable ethical costs by offering useful information to benefit future generations of patients. While the commercial interests of sponsors are understandable and inevitable, investigators and physicians should avoid any bias in planning a clinical trial, in its conduct, and in final analysis and publication.

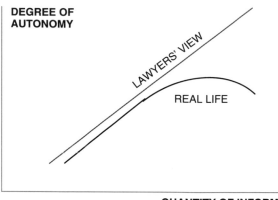

QUANTITY OF INFORMATION

Up to a certain point, understanding as a basis for autonomy increases with the amount of information. However, an overload of information obscures the essential facts and may lead to a decline in
understanding and in autonomy

I am writing this chapter with the experience of an oncologist who has spent the last
two decades working on lung cancer. I, therefore, borrow concrete examples from
this domain. However, similar ethical dilemmas in other fields of oncology and of
medicine could certainly be found.

In oncology, we are so often disappointed with the course of cancer treatment
that it would be unethical not to seek new, more efficient, safer, but also more affordable forms of diagnostics and treatment. The question is, therefore, not whether it is
ethical to do research but how to conduct research in such a way that the experience
will truly benefit future generations of patients. In other words: the foundation of
clinical research ethics is to seek truth; anything that misleads us or leads us astray
from this goal is unethical.

We will leave aside laboratory and epidemiological research and instead focus
on clinical research, that is, studies requiring the physical participation of patients,
which also raises the large majority of ethical concerns. We will begin our discussion of clinical research ethics with a short outline of ethical reflections on the first,
second, third, and fourth phases of clinical trials. We will then focus on issues
related to the patient's consent. Whether or not to obtain written consent for participation is no longer a question. While signed formal consent is nowadays a routine procedure, the real question is whether patients, in fact, understand the
information as the legal and ethical basis of consent.

A critical perspective on academic research and research funded by commercial sponsors reveals the not-so-rare cases in which personal or commercial
interests override the search for truth. The last part of the chapter offers a discussion on explicit and implicit forms of misleading research and fraud in clinical
trials.

19.1   The Four Phases of Clinical Trials

For a reader who may not have in-depth knowledge of medicine and medical research, we will start with a brief (and with some acceptable simplification) introduction to the four clinical phases of clinical trials. While we describe trials dealing with new drugs, a similar approach could well be applied to other novel modes of treatment.

In Phase I trials, a drug that was initially tested on animals is for the first time applied to humans. The goal is to determine the most appropriate dose and mode of administration, the pharmacokinetics and metabolism of the drug, and its main side effects. Of course, we are also monitoring the efficacy of the new drug against a specific disease, such as cancer, but this is not the primary objective of Phase I clinical trials. Recruited to Phase I clinical trials are either healthy volunteers or patients who cannot be offered any proven treatment, often with a broad spectrum of diagnoses. Phase I trials typically include a few dozen subjects.

Clinical research on a drug that passed the Phase I trial and therefore did not show an unacceptable level of toxicity continues with a Phase II clinical trial. When compared to Phase I, eligibility criteria for Phase II clinical trials are much narrower: the disease category is chosen after theoretical consideration on the potential efficacy of the drug against a particular disease, early anecdotal experience, and also on the basis of commercial interests. The primary objective of Phase II clinical trials is an assessment of drug efficacy. In cancer patients, drug efficacy is assessed by recording the percentage of patients with a significant reduction in cancer volume (assessment of remission), the median time to progression of the disease, and median survival. While monitoring the toxicity of the new treatment is mandatory, assessment of the quality of life is rarely reported. Phase II clinical trials can include anything from ten and up to one or two hundred patients.

If the outcomes of a Phase II clinical trial are encouraging, testing continues in Phase III, in which the new treatment is compared to the standard treatment in a randomized clinical trial. Since differences in efficacy are often small, confirmation of a statistically significant difference between treatment groups requires a large number of patients—anything from several hundred and up to more than a thousand. To recruit the required number of patients, Phase III clinical trials are usually conducted in collaboration between researchers from several hospitals

A drug that was proven advantageous over standard treatment in a Phase III clinical trial is registered by a national or international regulatory agency, such as the European Medicines Agency (EMA) or the Food and Drug Administration (FDA). After registration, the pharmaceutical company puts the drug on the market. Even after registration, the efficacy and toxicity of the drug are monitored in Phase IV clinical trials. Particular attention is paid to efficacy and to toxicity, as well as drug interactions in patient populations that have been poorly represented in earlier phases of clinical trials. Examples of such groups include the elderly, children, and patients with accompanying chronic diseases.

What are the main ethical concerns in each phase of clinical trials? In Phase I, it is crucial to clearly distinguish between drugs that do not affect the processes of cell division and anti-cancer drugs. The latter act on the processes of cell division and can, thereby, even in very small doses, lead to the development of new cancers.[1] Compared to other drugs, for which the participation of healthy volunteers is a generally accepted practice, cancer drugs are, therefore, never tested on healthy volunteers, but only in patients for whom there is no standard effective treatment available. The primary ethical concern pertains to the discord between expectations of patients and the probability of remission of cancer. The patient agrees to participate in the study because he expects the new drug to be beneficial; however, realistic chances for positive effects of treatment are below 5%. Such a low percentage is due to the fact that in a Phase I trial, a single drug, initially in very small doses, is applied to patients who have become resistant to combinations of multiple drugs.

When a Phase II clinical trial is planned, performed, and reported accurately and without biases (to be discussed shortly), there are no significant ethical concerns. The patient's expectations are in line with those of the physician: the patient hopes for the best treatment outcome, an objective that will also help the physician in the promotion of his research. Relations change in Phase III randomized clinical trials. The first and essential step is the unbiased randomization of patients, which should not be the responsibility of the physician involved in the trial. Unacceptable are procedures that would allow the physician to know in advance which treatment would be assigned to the next patient, for example, on the basis of odd and even numbers or on the date or day of the week. Prior to registration for the trial, neither the physician nor the patient should know the treatment that will be randomly assigned to the patient. For patient recruitment, the central ethical requirement is *equipoise*—a sincere uncertainty regarding the advantages of one or the other treatment. Therefore, if the physician proposes participation in a randomized trial, he may do so only when he is genuinely uncertain about the preferred treatment. The rule of genuine uncertainty applies not only to an entire group of patients with a specific diagnosis, but also every single patient. Whenever the physician feels that one treatment is preferable for a particular patient, he should not propose randomization.

Let me highlight another circumstance: genuine uncertainty tends to diminish over time. When the study is initially presented and approved, uncertainty is, of course, the basic condition for a random choice of treatment. Over the course of months and years, however, researchers can already anticipate which mode of treatment is better on the basis of intermediate study results and information available from other similar studies. Let us offer a concrete example: during the first 2 years since the beginning of a trial, 40 out of a planned total of 160 patients were randomized. Interim analysis reveals that among 20 patients in Group A, 14 are alive, whereas among 20 patients in Group B, only 6 are still alive. Even if the difference is not statistically significant, we can no longer talk about the genuine uncertainty

---

[1] Development of a second cancer five or more years after successful treatment of the first cancer is a significant problem in oncology and is especially worrying in pediatric oncology.

of the physician. The following question serves as the "golden rule": if the next patient were your sister, would you recommend a random choice of treatment? This leads us to the following conclusion: one of the ethical conditions in clinical trials with a random choice of treatment is the requirement that the period of participant recruitment be very short. This is necessary if we wish to avoid ongoing recruitment of new patients when the difference in the efficacy of the different modes of treatment is already obvious but has not yet reached the threshold of statistical significance.

19.2 Patient Information

Gone are the days when oral information and consent sufficed for participation in clinical trials. Today, written information and written informed consent is the applicable standard to all critical diagnostic and therapeutic interventions in everyday medical practice, and even more so to clinical trials.

The most frequent misleading practice nowadays is to provide for too extensive patient information written in a language which is very lengthy and barely understandable to persons without medical education. To take a concrete example, we provide here an excerpt from information offered to patients invited to a recent trial.[2] For a start, please note the highly cryptic title of the study:

> XX 123456—Randomized, open phase 3 testing of drug A alone or in combination with drug B in patients with advanced non-small-cell lung carcinoma with non-adenocarcinoma histology.

From the detailed 35-page information brochure, we provide an excerpt describing the process of blood sampling for the purposes of the study:

> Blood samples for determining the plasma concentration of drug B will be taken in cycle 1 within 2 hours before the infusion of drug B on day 1 and 1 hour after the infusion on day 2. In subsequent cycles, your PK blood samples will be taken within 2 hours before the infusion of drug B in cycles 2, 4, 5, and 6; and 1 hour after the infusion with drug B in cycle 5. If you have been randomly selected for administration of drug A only, but you are receiving drug B as addition to your best supportive care (BSC) after disease progression, they will take your PK samples within 2 hours before the infusion of drug B in cycles 1, 2, and 4.
>
> In patients that were randomly assigned to treatment with drugs A and B, serum samples for measurement of antibodies against the drug will be taken within 2 hours before the infusion of drug B in cycles 1, 2, and 4. In patients randomly assigned to treatment with drug A only but who receive drug B as addition to their "best supportive care" (BSC) after disease progression, samples for antibodies against the drug will be within 2 hours before the infusion of drug B in cycles 1, 2, and 4.
>
> During your visit upon the conclusion of treatment, blood samples for testing on IGF-IR positive CTCs will be taken as well…

---

[2] The data about the used drugs are not disclosed in order to prevent identification of the study and its sponsor.

How should an average patient understand 35 pages of such a text? Despite all the details, some crucial information is missing:

- How many times will the blood be taken exclusively for the purposes of the study (i.e., in addition to blood samples needed for oncological treatment)?
- What will be the total amount of blood taken?
- Will the patient have to remain hospitalized longer due to additional blood samples being taken?

In the last 20 years, the average length of written patient information in clinical trials has doubled [1]. It is no secret that such skewed "informing" of patients is due to the inappropriate involvement of legal experts in medical practice. The text is not written to better inform the patient, but to legally protect the study sponsor against potential compensation lawsuits. My recommendation to the legal experts and also physicians, who are after all responsible for medical research, is to use a bit of common sense. Common sense tells us that understanding will only increase to a certain reasonable length of the text, whereas with longer and longer texts, patient's understanding—and thus autonomy—decrease. Put differently: patients can be deceived by too little or too much information. The degree of deceit can become even worse when the study involves patients belonging to lower social classes, for whom it is even more difficult to expect that they will understand these extended texts.

What practical advice can we offer to improve informing patients? First, we should require that any patient information brochure exceeding five pages in length should contain a short, maximum one-page summary. This is the strategy we have accepted in all non-literary texts. The patient will read the summary and will create a mental schema to which he will attach the specific parts of the rest of the document.

Second, patient information should be subject to a quality control similar to that in regular use for all other diagnostic and therapeutic procedures. Understanding and remembering the crucial aspects of a trial should be tested on a sample of patients not only when informed consent is signed but also later during the course of the study. Such a procedure might well confirm the thesis that overly extensive information does not contribute to the patient's greater understanding of the study.

19.3   Academic Clinical Trials

It is my firm belief that academic clinical trials are necessary if we want to meet two objectives: one, to make progress in commercially less attractive areas of medicine and, two, to safeguard against a situation in which research is based exclusively on commercial interests. However, despite my support of academic research, I have to point out the potential biases.

What drives physicians to undertake clinical research even when they have no funding? There are two possible answers. First, it could be their sincere desire to explore new and potentially more successful, safer, more reliable, or more affordable modes of diagnosis and treatment. The second motivation is the desire for personal promotion: presentation of experience from a clinical trial often enables the researcher to advance in his career and to become known to his colleagues at home and abroad. There is nothing wrong with the second motivation as long as it supports the first one: advancement in one's career must unconditionally be founded on the search for objective truth.

Should academic studies be required to satisfy equally strict standards in all procedures as mandatory in studies with commercial sponsors? In some aspects, the answer is undoubtedly affirmative: a written protocol and ethical approval of the medical ethics committee, the respect of inclusion and exclusion criteria, comprehensible patient information, patient's written consent, and patient registration forms are equally important regardless of whether it is an academic study or a study with a commercial sponsor. The study protocol should also include a certificate of an insurance company. A decision regarding time-consuming copying of all diagnostics and of details of the treatment from the original documentation to the special clinical record forms (CRF) depends on the trial organization and may be unnecessary in trials performed only in a single institution. In academic clinical trials with no commercial sponsor, drugs that have already been registered and marketed are often used, in which case the requirement that all drugs used in the study should be wrapped in distinct packaging and clearly marked is redundant.

The most common form of bias, or sometimes overt fraud, in academic studies stems from ill-defined procedures of patient registration. One well-known example of such fraud was the Bezwoda affair. In a report on intensive therapy and bone marrow transplantation for patients with breast carcinoma, a South African researcher included data from fictive patients and from patients that did not meet the inclusion criteria [2]. There are rumors about a similar manipulation as the explanation for an exceptionally favorable report on the highly intensive treatment of lung cancer patients in several clinical trials from Kragujevac, Serbia [3]. If the process of registration is left to the researcher and no mechanisms of external control are in place, he can later exclude patients who did not respond to treatment and those with severe toxicity. Since such fraud may be difficult to prove, it is essential to prevent such behavior and entrust the registration of patients to a person who is entirely independent of the researcher.

The second significant bias in academic clinical trials is the non-publication of those clinical trials that did not show the expected positive results (*publication bias*). We do not have exact estimates about the scale of this phenomenon; we can, however, make indirect judgments by comparing scientific presentations at conferences where most authors present their research for the first time and the subsequent publication of articles in scientific journals, a task that requires considerably more work from the author. Such a comparison shows that a significant proportion of "negative" studies are never published as a full report in a scientific journal.

19.4   Clinical Trials with a Commercial Sponsor

Clinical research now inevitably depends upon support from the pharmaceutical industry. New drugs come to the market after the collaboration of basic biomedical research, the pharmaceutical industry, and physicians.

In their attempts to justify the high prices of new drugs, pharmaceutical companies often overestimate their role in research. Many key breakthroughs were not made by researchers in the R&D departments of pharmaceutical companies but by academic researchers seeking new ways of treatment or linking observations across different areas of medicine and biology. The three typical examples of such a way to discovery are penicillin as the first antibiotic, the first cytotoxic drugs for cancer treatment, and targeted anti-cancer drugs. Well-known is the story of Alexander Fleming and his discovery of penicillin. Fewer physicians know, however, that Sydney Farber carried out the first research on aminopterin (the first anti-cancer drug) without any financial support and against strong opposition from his colleagues and from potential sponsors. Half a century later, when cytotoxic anti-cancer drugs were widely used, the proponents of the idea of targeted anti-cancer drugs trastuzumab (nowadays a standard drug in breast cancer treatment) and imatinib (the most effective drug for gastrointestinal stromal tumors and for chronic myeloid leukemia) faced a similar atmosphere of scientific and commercial skepticism. The first clinical trials with these two medicines were performed despite the strong reluctance of pharmaceutical companies to share samples of drugs that were not yet registered and considered of no commercial interest [4]. While exceptions from this statement certainly exist, we may nevertheless say that many pharmaceutical companies are not very innovative in basic research and focus on the less risky endeavor of upgrading a promising finding from academic research into a drug that will be registered and will obviously bring profits.

It is not the purpose of this chapter to deny or diminish the essential and often crucial role of sponsors, especially of the pharmaceutical industry, in their financial and logistic support of clinical trials. Rather, we wish to point to the fact that financial interests may lead to a temptation to manipulate the planning, performance, analysis, and reporting of results of clinical research. Recognizing a potential bias is of importance to every physician, whether a researcher, an opinion leader, an editor or reviewer of a paper submitted to publication, a member of a reimbursement committee, or a clinician who will consider whether to adopt the new experience into his clinical practice.

Planning a clinical trial is not pure science. Every sponsor has a legitimate interest to choose a design which will most likely lead to the final goal—promotion of a new treatment. Unfortunately, however, honesty often succumbs to financial interests, leading to a biased design of a trial. In trials with a random choice of treatment, the control group of patients is often offered a treatment that is inferior to the best standard treatment. Let us consider a concrete example of a large randomized trial for patients with pleural mesothelioma. In this trial, patients treated with a new drug, pemetrexed, in combination with cisplatin were compared to the control group that received monotherapy with cisplatin [5]. However, over the last 15 years, cisplatin as monotherapy has not been used for this indication since several other combinations of drugs were clearly superior to cisplatin alone. For regulatory agencies, such a biased design was not considered an issue, and on the basis of experience from a trial with a suboptimal control group, pemetrexed was registered for treatment of mesothelioma. Thus, an expensive new drug is now in routine use for patients, most of whom are low-income asbestos workers. In contrast to this study, our thoracic oncology team used an alternative combination of drugs (low-dose gemcitabine in combination with cisplatin) and completed a trial that led to very favorable results, while the costs for the drugs were less than one tenth of those in the aforementioned trial with pemetrexed [6]. It comes as no surprise that no sponsor was interested in offering support to our trial.

Many readers will think that a biased choice of control group is a rare exception. Unfortunately, this is not the case: acceptability of systematic bias appears even in such an important document as the most recent version of the Declaration of Helsinki—Ethical Principles for Medical Research Involving Human Subjects of the World Medical Association. In Article 33 of the document, one can read the rules for selecting the control group: "The benefits, burdens and effectiveness of a new intervention must be tested against those of the best proven intervention(s), except [...] where for compelling and scientifically sound methodological reasons the use of any intervention less effective than the best proven one, the use of placebo, or no intervention is necessary to determine the efficacy or safety of an intervention" [7].

Therefore, we know what the best-proven treatment for a specific disease is; however, we are planning a study in which half of the patients will receive a new, unproven treatment and the other half a placebo or a treatment that is inferior to the standard treatment. We may be confident that the fact of suboptimal treatment for patients in the control group will not be disclosed in patient information documentation, which of course means that also the patient's consent for participation in the study loses its ethical and legal validity. It is hard to imagine a scientifically justified reason for performing a randomized clinical trial that does not take the best-proven treatment as a baseline, and it is even more difficult to understand how such a text could have been approved by the World Medical Association. However, if that were their decision, it would also be fair if they took a step forward and replace "scientifically sound methodological reasons" with "commercially sound reasons."

The definition of *trial objectives* can also be biased. From a patient's perspective, only two objectives are of importance: to live better, or to live longer; and if possible, better and longer. Translated into scientific language, quality of life and survival should be the primary objectives. All the rest, such as a temporary radiological reduction of the tumor (remission of the disease) or time to progression of disease are of minor interest to patients and should be analyzed as secondary objectives. Of what benefit to a patient is a brief remission if accompanied by severe toxicity, or longer time to progression at the expense of numerous additional visits to a hospital and no improvement in survival? These—for the patient crucial—aspects are often ignored in pharmaceutical studies in which secondary objectives are listed as the primary ones, while the quality of life is rarely recorded. Let me provide the findings of my own review of clinical trials on treatment of advanced lung cancer, a disease with which only a small fraction of patients survives more than 2 years after the diagnosis and cure is extremely rare. Among 349 studies published between 2013 and 2015, only 21% included quality of life among trial objectives [8].

According to *inclusion and exclusion criteria* for selection of patients for a trial, often only patients in excellent general condition without accompanying diseases are selected for participation. Both sponsors and researchers themselves are responsible for such a selection since they are reluctant to enroll patients with a higher risk of adverse events. From the sponsor's perspective, this attitude is understandable since the enrollment of patients with co-morbidity and receiving other medications may render the final analysis of the trial quite difficult. However, physicians themselves are also responsible for offering a trial only to patients in excellent general condition. An experienced investigator knows that every case of a serious adverse event brings significant additional work for reporting and for complications and many post hoc explanations. However, if the patient population is limited to those in the best general condition, then the reported experience has only limited validity for all other patients who may be elderly, frail, or with co-morbidity.

Performance of a trial can also contribute to biases. When the primary goal of the study is overall survival, the procedure for both groups must be comparable across all relevant factors. In the case of progression after primary treatment, most trial protocols do not specify the precise procedure for second-line treatment, which may significantly influence survival. A typical example is a study on maintenance treatment in patients with advanced lung cancer. In this trial, patients with advanced lung cancer and treated with first-line chemotherapy were randomized between those who did or did not receive maintenance with pemetrexed [9]. Where do we see a bias? At the time of disease progression, patients in the control group should be offered what the first group received immediately after first-line treatment. However, only 15% of patients in the control group received any form of second-line treatment. Although the experience from this trial is interpreted as evidence in favor of maintenance treatment, the question of whether immediate maintenance treatment with pemetrexed is superior to the use of this drug as a second-line therapy remains unresolved.

Analysis and publication of results are entirely in the hands of the sponsor. While the first author is an academic researcher, the analysis and interpretation of the trial experience are often prepared by the sponsor. Along with emphasis of the positive findings, observations that do not support the introduction of a new treatment are offered much less attention, and comments on price and affordability of a new treatment are never included in the final publication. As in academic research, sponsored studies with negative results are often not published or are published in a journal with a low impact factor (*publication bias*).

19.5   Progress, Patient Solidarity, and Honesty

After all these critical words about clinical trials, I should stress that there can be no progress without research. It is a matter of thousands of small steps and every once in a while, a significant leap forward. Research not only benefits future patients, but also those who participate in the study. In a Phase III clinical trial, even patients who were included in the group receiving standard treatment have better chances than patients who received the same treatment outside the clinical trial. This observation can be explained by the fact that in clinical trials all diagnostic and treatment procedures are very precisely determined and controlled. Clinical research therefore indirectly improves the quality of our routine medical work.

At the same time, we cannot ignore the ethical burden of clinical research. In Phase I clinical trials, we discussed the gap between the patient's expectations and realistic expectations about disease improvement. Despite the most honest and careful design, conduct, analysis, and reporting possible, certain ethical concerns are unavoidable, especially in randomized Phase III trials. Such ethical burdens can be justified by assuming inter-generational solidarity among patients. Let me explain: the patient of today expects the physician to recommend the preferred treatment. To do so, the physician will present results of recent clinical trials. Yesterday's patients, therefore, help in the search for the best treatment for today's patients. Today's patients, in turn, will help us advance our knowledge even a small step further, to benefit tomorrow's patients.

It is essential not to lose our sight of two objectives: honesty towards patients participating in clinical research, and honesty towards society as a whole as well as to future patients who should be offered objective information. In terms of honesty towards patients, the key is good communication, including comprehensible and understandable information about the clinical trial. As far as honesty towards society is concerned, we as researchers and as physicians must persist in our search for truth. We must, therefore, go beyond our personal ambitions or the interests of sponsors and present the experience in an unbiased manner.

References

1. Berger O, Grønberg BH, Sand K, Kaasa S, Loge JH. The length of consent documents in oncological trials is doubled in twenty years. Ann Oncol. 2009;20:379–85.
2. Droste S, Herrmann-Frank A, Scheibler F, Krones T. Ethical issues in autologous stem cell transplantation (ASCT) in advanced breast cancer: a systematic literature review. BMC Med Ethics. 2011;12:6.
3. Jeremic B, Milicic B, Milisavljevic S. Clinical prognostic factors in patients with locally advanced (stage III) nonsmall cell lung cancer treated with hyperfractionated radiation therapy with and without concurrent chemotherapy: single-institution experience in 600 patients. Cancer. 2011;117:2995–3003.
4. Mukherjee S. The emperor of all maladies: a biography of cancer. New York: Scribner; 2010. ISBN: 978-1-4391-0795-9.
5. Vogelzang NJ, Rusthoven JJ, Symanowski J, Denham C, Kaukel E, Ruffie P, et al. Phase III study of pemetrexed in combination with cisplatin versus cisplatin alone in patients with malignant pleural mesothelioma. J Clin Oncol. 2003;21:2629–30.
6. Kovac V, Zwitter M, Rajer M, Marin A, Debeljak A, Smrdel U, et al. A phase II trial of low-dose gemcitabine in prolonged infusion and cisplatin for malignant pleural mesothelioma. Anti-Cancer Drugs. 2012;23:230–8.
7. https://www.wma.net/wp-content/uploads/2016/11/DoH-Oct2013-JAMA.pdf
8. Zwitter M. Toxicity and quality of life in published clinical trials for advanced lung cancer. Support Care Cancer. 2018;26:3453–9. https://doi.org/10.1007/s00520-018-4214-1.
9. Ciuleanu T, Brodowicz T, Zielinski C, Kim JH, Krzakowski M, Laack E, et al. Maintenance pemetrexed plus best supportive care versus placebo plus best supportive care for non-small-cell lung cancer: a randomised, double-blind, phase 3 study. Lancet. 2009;374: 1432–40.

Suggested Reading

Amsterdam JD, McHenry LB, Jureidini JN. Industry-corrupted psychiatric trials. Psychiatr Pol. 2017;51:993–1008. https://doi.org/10.12740/PP/80136.

Bell JA, Balneaves LG. Cancer patient decision making related to clinical trial participation: an integrative review with implications for patients' relational autonomy. Support Care Cancer. 2015;23:1169–96. https://doi.org/10.1007/s00520-014-2581-9.

Denson AC, Mahipal A. Participation of the elderly population in clinical trials: barriers and solutions. Cancer Control. 2014;21:209–14.

Every-Palmer S, Howick J. How evidence-based medicine is failing due to biased trials and selective publication. J Eval Clin Pract. 2014;20:908–14. https://doi.org/10.1111/jep.12147.

Flacco ME, Manzoli L, Boccia S, Capasso L, Aleksovska K, Rosso A, et al. Head-to-head randomized trials are mostly industry sponsored and almost always favor the industry sponsor. J Clin Epidemiol. 2015;68:811–20. https://doi.org/10.1016/j.jclinepi.2014.12.016.

Flory JH, Mushlin AI, Goodman ZI. Proposals to conduct randomized controlled trials without informed consent: a narrative review. J Gen Intern Med. 2016;31:1511–8. PMID: 27384536.

Galton DJ, Dodge JA. Legal distinctions between clinical research and clinical investigation: lessons from a professional misconduct trial. Int J Philos Stud. 2013;1:13–6.

Gillies K, Entwistle V, Treweek SP, Fraser C, Williamson PR, Campbell MK. Evaluation of interventions for informed consent for randomised controlled trials (ELICIT): protocol for a systematic review of the literature and identification of a core outcome set using a Delphi survey. Trials. 2015;16:484. https://doi.org/10.1186/s13063-015-1011-8.

Glackin SN. Placebo treatments, informed consent and 'the grip of a false picture'. J Med Ethics. 2015;41:669–72. https://doi.org/10.1136/medethics-2014-102332.

Halkoaho A, Pietilä AM, Ebbesen M, Karki S, Kangasniemi M. Cultural aspects related to informed consent in health research: a systematic review. Nurs Ethics. 2016;23:698–712. https://doi.org/10.1177/0969733015579312.

Herson J. Strategies for dealing with fraud in clinical trials. Int J Clin Oncol. 2016;21:22–7. https://doi.org/10.1007/s10147-015-0876-6.

Kuthning M, Hundt F. Aspects of vulnerable patients and informed consent in clinical trials. Ger Med Sci. 2013;11:Doc03. https://doi.org/10.3205/000171.

Lamont S, Jeon YH, Chiarella M. Assessing patient capacity to consent to treatment: an integrative review of instruments and tools. J Clin Nurs. 2013;22:2387–403. https://doi.org/10.1111/jocn.12215.

Linker A, Yang A, Roper N, Whitaker E, Korenstein D. Impact of industry collaboration on randomised controlled trials in oncology. Eur J Cancer. 2017;72:71–7. https://doi.org/10.1016/j.ejca.2016.11.005.

Leibson T, Koren G. Informed consent in pediatric research. Paediatr Drugs. 2015;17:5–11. https://doi.org/10.1007/s40272-014-0108-y.

Liao SM, O'Neil C, editors. Current controversies in bioethics. Abingdon: Routledge; 2017. ISBN-13: 978-1138855823.

Myles PS, Williamson E, Oakley J, Forbes A. Ethical and scientific considerations for patient enrollment into concurrent clinical trials. Trials. 2014;15:470. https://doi.org/10.1186/1745-6215-15-470.

Nishimura A, Carey J, Erwin PJ, Tilburt JC, Murad MH, McCormick JB. Improving understanding in the research informed consent process: a systematic review of 54 interventions tested in randomized control trials. BMC Med Ethics. 2013;14:28. https://doi.org/10.1186/1472-6939-14-28.

Probst P, Knebel P, Grummich K, Tenckhoff S, Ulrich A, Büchler MW, Diener MK. Industry bias in randomized controlled trials in general and abdominal surgery: an empirical study. Ann Surg. 2016;264:87–92. https://doi.org/10.1097/SLA.0000000000001372.

Sture J. The ethics and biosecurity toolkit for scientists. Singapore: World Scientific; 2016. ISBN-13: 978-1786340924.

Williams CJ, Zwitter M. Informed consent in European multicentre randomised clinical trials—are patients really informed? Eur J Cancer. 1994;30A:907–10.

Willmott C, Macip S. Where science and ethics meet: dilemmas at the Frontiers of medicine and biology. Westport, CT: Praeger; 2016. ISBN-13: 978-1440851346.

Zaner RM. A critical examination of ethics in health care and biomedical research: voices and visions. Berlin: Springer; 2016. ISBN-13: 978-3319382517

Zhang D, Freemantle N, Cheng KK. Are randomized trials conducted in China or India biased? A comparative empirical analysis. J Clin Epidemiol. 2011;64:90–5. https://doi.org/10.1016/j.jclinepi.2010.02.010.

Zwitter M. A personal critique: evidence-based medicine, methodology, and ethics of randomised clinical trials. Crit Rev Oncol Hematol. 2001;40:125–30.

Zwitter M, Tobias JS. A survey of the ethical considerations in randomised trials for lung cancer. Lung Cancer. 1998;19:197–210.

Unproven Methods of Diagnostics and Treatment

20

Abstract

Labels such as holistic approach or integrative medicine are often used to emphasize differences between conventional medicine and unconventional methods of diagnostics and treatment. Yet, such a distinction is not justified as it implies that conventional medicine does not serve the physical, psychological, and spiritual needs of our patients. Both from the scientific and the ethical standpoint, unproven methods of diagnostics are unacceptable. Among therapeutic measures, many complementary methods may benefit the patient. On the other hand, unproven procedures as an alternative to standard treatment of proven efficacy often lead to deterioration of patient's health and should not be applied by persons with a license of a medical doctor. A similar critical attitude is appropriate also for drugs which are still in an early stage of development. In this case, a clinical trial is the only acceptable option. If we really follow the interests of patients and not those of the pharmaceutical industry, the *Right to Try* legislation which allows the early use of unapproved drugs for critically ill patients with no standard treatment left is not acceptable.

Extracts from European Mistletoe (*Viscum album*) are among the most popular alternative drugs against cancer. Their use continues despite the fact that among dozens of trials, not a single one has provided clear evidence for efficacy [1, 2]

Medical faculties do not hold a monopoly on everything physicians do in the course of their profession, even less so over what persons without medical education do in the broader field of healthcare. The sources of unproven modes of diagnostics and treatment are extremely diverse: traditional folk and herbal medicine (local and global), present-day non-scientific approaches to human biology and physiology, but also a pharmaceutical industry that markets products even before they pass all the required market approval procedures. Much in the same way that any treatment offered within "official" medicine must be critically evaluated; methods offered outside official medicine are not all bad. As we wrote in the Code of Medical Ethics, the main dividing line falls between acceptable *supplementary* measures and unproven *alternative* modes of treatment, which are very problematic. Unproven modes of treatment are frequently acceptable, often even useful if they supplement standard modes of treatment. It becomes dangerous for the patients, and hence unacceptable when standard modes of treatment are replaced with wizardry.

A comprehensive view of the patient and the disease, holistic approach, integrative medicine—these are but some of the labels that supposedly support diagnostic and therapeutic procedures that are not scientifically tested and are not endorsed by the official medicine. With such sweet talk, they are forcing upon us the idea that official medicine does not treat patients holistically—an assumption that has no grounds. Physicians who follow the guidelines for good communication (the topic of Chap. 8) and perform their work honestly and with compassion treat patients far more holistically than many of the advocates of alternative medicine do. Physicians who are not devoted to their patients and primarily follow their personal interests are too common in both camps: insider official medicine and outside it.

20.1   Alternative Diagnostics

Energy systems of the body, bioresonance, karmic diagnostics, visual diagnostics, pulse diagnostics, iridology—this is just a short list of alternative techniques with which one could supposedly detect a disease and establish a diagnosis. At first glance, alternative diagnostics is of marginal importance since the practitioners are not performing any treatment and therefore cannot harm the patients' health. However, the problem is much greater than it seems because people are in fact paying for useless "test results." This is more than just fraud. Concerned individuals bring these "results" to their personal physician and then demand that he carry out "official" diagnostics. The physician is faced with the dilemma: even though there is no reasonable indication for performing a computed tomography (CT) scan, it is, of course, possible that after a year, one out of 50 patients will, in fact, develop cancer. One can only imagine what the reaction of the patient, the journalists, and the public would be in such a scenario. Karmic diagnostics will be uncritically acclaimed, and the physician will face an avalanche of accusations about how he arrogantly disregarded allegedly clear evidence of disease and is now responsible for the life of the patient. During the next similar event, the experience of media lynching may lead to a different decision: the physician may disregard professional guidelines and follow the demand for additional diagnostics. It goes without saying that this means irrational spending of public healthcare funds and increased waiting periods.

Alternative diagnostics can instill unfounded fears, but they can also generate a false impression of safety. Here is an invitation to alternative diagnostics: "A comprehensive examination of the electromagnetic fields over the entire body (without the need to undress and without harmful radiation) precisely indicates any changes that signify disease." The text is taken from an advertisement by a private physician. I should explain to our non-medical readers that for every diagnostic method we need to determine the probability of detecting an already present disease (sensitivity) and the probability for correct identification of the disease (specificity). Let us take early breast cancer detection as an example for which techniques such as mammography, ultrasound breast examination, and magnetic resonance imaging are standard methods with well-known sensitivity and specificity profiles. A comprehensive examination of the body's electromagnetic fields certainly fares much worse in all respects, which means some women would end up paying for diagnostics that will not warn them of an already present disease.

My opinion is clear. A physician who holds a degree of medical doctor and has a license of the medical chamber should not practice alternative diagnostics.

20.2   Supplementary and Alternative Treatment

Conflicting opinions are heard about whether or not it is acceptable for the physician to perform unproven modes of treatment. Before we explore in more detail the ethical status of unproven modes of treatment, let us consider the following story told by a German teacher[1]:

> In a primary school class, one of the boys was extremely restless. He could not control himself; he was disrupting the class; he would stand up in the middle of a lecture and walk to the other side of the room. The school physician did not give any medications; instead, she gave him a pendant with a green stone and told him to hold the stone whenever he felts the urge to stand up. From that moment onward, the boy has been causing no more trouble, and he himself is happy that he can calmly follow the class.

Parallel to this story are the data on the heavy and increasing use of sedatives, especially in underage persons. Slovenia has a population of 2.07 million of which 290,000 inhabitants are in the 5–19-year-old age group. In 2009, physicians issued 5440 prescriptions for anxiolytics and antidepressants to those younger than 19 years of age. In 2014, the number rose to over 8400 prescriptions issued [3]. Why am I reporting the data on the use of medications for relieving minor aberrations in behavior along with the story above? I wish to stress that it is not always the best strategy if physicians immediately resort to prescribing drugs offered by official medicine. Many troubles can be alleviated by a long conversation, traditional herbal medicine, or by offering advice about lifestyle.

With unproven modes of treatment, I nevertheless advise physicians to be particularly cautious. The physician who approaches the patient with the authority granted by the degree of a medical doctor cannot and should not deny the fundamentals of scientific medicine. He will use homeopathy, herbs, or other unproven methods only in cases when there is no indication for treatment with professionally and scientifically accepted modes of treatment; even in those cases, however, he will not exploit the patient to his own financial interest. Someone with a medical diploma and license, therefore, knows that there is no convincing double-blind study that would confirm the efficacy of homeopathic medications and that homeopathy therefore only works as a placebo. Those physicians who "turned their coats" and have become believers in alternative procedures for diagnostics and treatment should return their medical licenses and clearly inform their patients about it. A physician who believes that he can treat malaria with homeopathic drops is dangerous for patients and therefore has no place in our profession.

20.3   Right to Try

Over the last 3 years, legislation on the possibility to treat the terminally ill with medications that have not yet obtained market approvals has emerged in the USA. The rationale behind this move was that terminally ill patients do not have time to

---

[1] Personal communication. January, 2016.

wait for the official approval of new drugs and, therefore, should be given the opportunity to try treatment with medications that are still in the testing phase. More than half of the federal states have already passed laws grouped under the label "right to try." Treatment with medications without market approval is allowed under the following conditions [4]:

- The disease is in the final stage, incurable, and there is no proven treatment left.
- The medication has passed Phase 1 clinical trials.
- A person responsible for treatment (who is not necessarily a physician) agrees to try the treatment.
- The patient or his guardian is informed and signs the consent for treatment with untested medication.
- In cases of complications, the drug manufacturer does not accept responsibility.

At first glance, it seems positive that the legislation at least gives hope to terminally ill patients. However, let us critically compare *the right to try* with a practice that has long been established globally and is known as the *compassionate use of new drugs*. They differ in three main respects: levels of safety and efficacy, oversight by the drug authority, and payment. *Compassionate use* is only possible once the experience with the new drug is already known from at least Phase 2 if not even Phase 3 clinical trials. At that point, it is only a matter of time for the drug to obtain international and national market approvals. *The right to try* brings the drug to the patient much earlier—after finalized Phase 1 clinical trials—when the first information about the toxicity of the drug is known, but experience regarding its efficacy is very limited. *Compassionate use* is overseen by the drug authority, whereas *the right to try* bypasses all means of control and depends on the direct contact between the pharmaceutical company and the patient or his guardians. Finally, the third important difference: payment. The drug in *compassionate use* is available free of charge; the pharmaceutical company sees this as a form of early marketing as sales will proceed quickly once the registration is finalized. Under *the right to try*, however, the manufacturer is allowed to sell the drug—even though the drug is less safe and we have no information about its efficacy.

I do not see any advantage in legislation that allows and encourages *the right to try*, and I hope the US practice will not spread to the other side of the Atlantic. I am afraid this practice opens up the way of exploiting false hope in order to profit from the most severely ill patients. We know that less than one in ten medications passes all three phases of clinical trials, with subsequent registration and regular use. Some companies might realize that their new drug will not pass all phases to registration and will be tempted to delay the publication of poor results in Phase 2 trials. Instead, through *the right to try* mechanism, they could nevertheless attempt to sell the drug without any oversight of the medical authorities. This, therefore, leads to an impression that *the right to try* brings much more benefit to pharmaceutical companies than to patients.

References

1. Melzer J, Iten F, Hostanska K, Saller R. Efficacy and safety of mistletoe preparations (*Viscum album*) for patients with cancer diseases. A systematic review. Forsch Komplementmed. 2009;16:217–26. https://doi.org/10.1159/000226249.
2. Evans M, Bryant S, Huntley AL, Feder G. Cancer Patients' experiences of using mistletoe (*Viscum album*): a qualitative systematic review and synthesis. J Altern Complement Med. 2016;22:134–44. https://doi.org/10.1089/acm.2015.0194.
3. Delo newspaper, Ljubljana, December 21, 2015.
4. Rubin MJ, Matthews KR. The impact of right to try laws on medical access in the United States. The Baker institute policy report. 2016;66.

Suggested Reading

Barić H, Đorđević V, Cerovečki I, Trkulja V. Complementary and alternative medicine treatments for generalized anxiety disorder: systematic review and meta-analysis of randomized controlled trials. Adv Ther. 2018;35:261–88. https://doi.org/10.1007/s12325-018-0680-6.

Bedlack RS, Joyce N, Carter GT, Paganoni S, Karam C. Complementary and alternative therapies in amyotrophic lateral sclerosis. Neurol Clin. 2015;33:909–36. https://doi.org/10.1016/j.ncl.2015.07.008.

Carrieri D, Peccatori FA, Boniolo G. The ethical plausibility of the 'Right to Try' laws. Crit Rev Oncol Hematol. 2018;122:64–71. https://doi.org/10.1016/j.critrevonc.2017.12.014.

Holbein ME, Berglund JP, Weatherwax K, Gerber DE, Adamo JE. Access to investigational drugs: FDA expanded access programs or "Right-to-Try" legislation? Clin Transl Sci. 2015;8:526–32. https://doi.org/10.1111/cts.12255.

Mullins-Owens H. Integrative health services: ethics, law, and policy for the new public health workforce. Berlin: Springer; 2016. ISBN-13: 978-3319298559.

Sandman L, Liliemark J. From evidence-based to hope-based medicine? Ethical aspects on conditional market authorization of and early access to new cancer drugs. Semin Cancer Biol. 2017;45:58–63. https://doi.org/10.1016/j.semcancer.2017.05.009.

Stuttaford M, Al Makhamreh S, Coomans F, Harrington J, Himonga C, Hundt GL. The right to traditional, complementary, and alternative health care. Glob Health Action. 2014;7:24121. https://doi.org/10.3402/gha.v7.24121.

Physicians Beyond Patient Care

21

Abstract

The last chapter deals with a physician who is not directly involved in patient care. Some physicians will advance to a leading position, either within a health-care system, in governmental structures or in pharmaceutical industry. Other situations include teaching responsibilities; the role of an expert in the court; medical reports for candidates for driving licenses or use of weapons; expert opinions on investments in medical equipment; and medical counselling in sports, especially those with strong commercial support. In all these situations, physicians are obliged to act honestly, in the best interests of the society as a whole. Numerous cases of misbehavior have been revealed; yet, it is hard to understand those of us who cross the line of professional integrity and benefi-cence. Whatever a physician is doing outside of his basic profession, honest behavior enables an eventual return to medical practice.

© Springer Nature Switzerland AG 2019
M. Zwitter, *Medical Ethics in Clinical Practice*,
https://doi.org/10.1007/978-3-030-00719-5_21

Scott Westgarth, 31, is the last on the list of more than 500 boxers who died in the ring or as a result of boxing [1]. In addition, many (probably most) boxers have long-term mental and neurological consequences due to chronic traumatic encephalopathy [2]. The British Medical Association, the American Academy of Pediatrics, and the Canadian Paediatric Society oppose boxing as a sport for children and adolescents [3, 4]

In this final chapter, I will highlight the position of the physician in situations in which his role is not to perform diagnostics or treatment of patients. It is essential that the physician and all others involved understand how interpersonal relations, rights, and responsibilities change in such situations.

21.1   Managerial Responsibilities and Politics

Some male physicians (rarely female physicians) move up to assume managerial positions during their professional careers. Many are led by their honest desire to improve the conditions and act for the benefit of patients and the entire healthcare staff. It is common to criticize leadership ambitions in advance; however, this is inappropriate especially if the criticisms are coming from those who never assume broader responsibilities themselves.

The physician who takes on a managerial position or even a political role should be aware that he must maintain honest contact with his colleagues in medical practice if he is to perform his new duties successfully. Although he will not be able to avoid all conflicts, he will nevertheless enjoy support as long as he directs his efforts to the benefit of his department or institution and maintains honest bidirectional communication with subordinate staff. When healthcare is constantly faced with shortages in funding, staff, equipment, and appropriate space, a good physician in a managerial position will constantly find himself in tense relations with his superiors—with the director-general or with the minister and the minister alongside the prime minister.

A reasonable leader knows that while managerial positions are granted for a mandate or two, the medical profession is for life; clinging to a leadership function against the will of the collective or even by pursuing lawsuits is indecent and can only lead to bitter feelings. I can assure the readers from my own experience that stepping down from a leadership function and returning to medical practice can bring relief, invoking feelings similar to when you take off heavy boots after a whole day of skiing.

21.2 Expert

The physician assumes the role of expert whenever his work is not meeting the demands of a patient, but those of a third party: the court system, police, public administration, insurance company, or employer. In a broader sense, the expert role also includes, for example, issuing a health certificate for proving the ability to drive motor vehicles or issuing a gun license.

The person who is the subject of expert opinion and the physician must realize that this is not a case of the physician–patient relationship in which the interest of the patient comes first. The physician-expert must clearly follow the questions and provide clear answers: no less and no more than what is being asked.

Expert duties also include certain responsibilities of physicians who otherwise work with patients. Issuing a certificate of absence from work due to sick leave is not part of diagnostics or treatment. Rather, this is a document issued by the physician for the needs of the employer or health insurance company, and the physician, therefore, is in the role of an expert. Many physicians are not aware of this distinction. A physician who would follow patient's best interest and issue a health certificate even when absence from work is not medically justified should understand that he may face charges of issuing a false certificate.

21.3 The Physician, the Pharmaceutical Industry, and Medical Equipment Providers

Some aspects of the relations with the pharmaceutical industry have been briefly discussed in the chapter on research.

We cannot and also do not want to prohibit professional contacts of physicians with the representatives of the pharmaceutical industry and manufacturers of medical equipment. Undoubtedly, in this way physicians gain important information about new medications and medical equipment for more precise diagnostics and more successful treatment. The support of pharmaceutical companies is also welcome, often necessary, for the organization of professional meetings. However, an overly strong dependence on or servility towards the representatives of the pharmaceutical industry or medical equipment providers could be a form of corruption.

To avoid allegations of corruption, I provide short recommendations here:

- Contacts with the representatives of pharmaceutical and other companies should be limited in time and must take place outside regular working hours. Any contacts should be announced in advance and communicated to the management of the health institution or hospital.
- Medications with which we have positive experience should not be replaced with newer drugs.
- The physician should never accept a monetary or material reward for prescribing certain medications. This includes the prohibition of collaborating in studies in which the physician receives a reward for prescribing a new drug and writing a short report.
- All donations from pharmaceutical and other companies for educational and other similar purposes must be addressed to the healthcare institution, never to the physician directly.
- In public professional presentations, the physician must declare potential conflicts of interest. A professional presentation must be balanced and impartial.

21.4   Teacher

Medicine cannot be learnt from books or professional articles alone. It is inevitable that written guidelines cannot list all the factors influencing physician's decisions. Books and articles, therefore, cannot replace the personal experience that is communicated to young people by a good teacher. This holds particularly true for relationships and communication with patients for which a personal example is indispensable.

All physicians should know that they are being observed by their younger colleagues and that their example is transferred to the next generation, for better or for worse. In the remainder of the chapter, I will focus on the teaching that some of us do in classes of medicine.

It is not possible to imagine lectures in medicine without the presence of patients. We somehow expect that the patients treated in university healthcare institutions will understand and also accept their roles in classes. However, our expectations must not be equated with the patient's consent: every patient has the right to refuse participation in classes. It is, therefore, not appropriate if the physician brings a group of students to the hospital room without announcing it in advance. Patient's privacy must be guaranteed to a reasonable extent, especially when gathering anamnesis or during physical examinations. Also important is the requirement to keep student groups small: it is not reasonable to expect that a patient will be comfortable with the idea of 12 students learning how to perform breast examination on her.

During practical classes, many questions on diagnostics and treatment arise. Students should understand that there will be time for questions and answers once they leave the hospital room. It is highly inappropriate if a discussion that could hurt the patient develops in his presence.

21.5 The Physician in Public

The physician should not glorify himself and his work in public—this is an established ethical recommendation. It is particularly inappropriate if the physician is boasting and comparing himself to other physicians.

Nevertheless, we cannot entirely bypass the public. In the chapter on preventive medicine, we noted that one of the duties of every physician is raising awareness of the public about the importance of leading a healthy lifestyle. The media and the internet are full of disinformation and utterly misleading advice, for example, on the rejection of vaccination or the alleged advantages of severely unbalanced diets. Physicians are left with no other strategy than to speak up and attempt to at least mitigate the damage caused by such disinformation. Physicians should react to sensationalistic reports. A cure for cancer is often presented as a miracle and attributed to alternative "treatment," ignoring the fact that the patient also received standard anti-cancer therapy. Similarly, early laboratory experience in a potential new treatment is interpreted as a break-through achievement of proven value. It is sad, and indeed unethical that many journalists and media outlets do not wish to see the harm caused by such unbalanced reporting. Finally, there are the reports on professionally questionable medical behavior. The media often accuse the physician even before the charge is filed and long before the judgment is final, frequently disclosing his full name. As I stated in the chapter on professional malpractice, it is the patient who delineates the field of confidentiality. When a patient presents his case to the public, the physician has all the right to clearly disclose his side of the story.

When it comes to public engagement, physicians have another essential task: we must inform the public about the flaws in the healthcare system. The belief that nothing will change without public pressure is not groundless, whether it pertains to the employment of young physicians or to the urgent replacement of a 15-year-old X-ray device.

21.6 Sports Medicine

Do sport and health walk together, hand in hand, or are they opponents?

There is no doubt that sport is a useful and efficient antidote to present-day stressful, unhealthy lifestyles and to all modern addictions—from alcohol and drugs to computer addictions. This is especially true for recreational sports and young people.

Wherever large amounts of money flow, greed conquers honesty. This holds true in politics, public investments, research, and, sadly, also in sports. The observation pertains not only to professional sports but also to recreational sports. The physician's responsibility is to offer advice about the diet, training, and prevention or treatment of injuries. The line between these activities and the use of illegal stimulants and all sorts of doping is thin, however. The excuse "everyone is doing it" is simply unacceptable: physicians should never dishonestly support the successes of his protégés and are responsible for the permanent consequences of doping on the health of the athletes.

Should the physician be involved in sports that have been proven to be harmful to health? Boxing is an example of a sport causing permanent brain damage through micro-injuries. This applies to junior boxers who wear boxing headgear as well; headgear only protects against external injuries, but it does not protect from injuries of the brain, which are the result of the impacts on the skull when blows are delivered. We are, therefore, not referring only to cases of severe facial injuries or even immediate death after the fight; in fact, injuries frequently lead to permanent disability. The British Medical Association and the World Medical Association are therefore calling for a ban on boxing and advise physicians to avoid active participation in that sport.[1]

21.7   Physician as Patient

From Hippocrates onwards, solidarity between physicians is featured in all documents on medical ethics. To our colleagues and also to other healthcare workers and students of medicine, physicians will devote particular attention and offer them priority treatment. In countries with waiting times for medical services, the profession is not among the legal criteria for assigning a priority in access to healthcare. Therefore, offering priority to our colleagues is an example of legislation and ethics being at cross-purposes. We can only hope there will not be a superintendent who will pick up on the supposed privileges of our profession. We will remind him that things have been that way for more than 2000 years; and will tell him that physicians take almost four times fewer days of sick leave than their patients would be approved for the same disease [5]. To speak of physician's privilege is therefore not only exaggerated but also deplorable.

When discussing the physician in the role of a patient, we must highlight three additional aspects. First, our sick colleague should receive the same thorough care as any other patient would. It may appear redundant to state so; however, it is unfortunately not. A sick physician also needs to have a detailed anamnesis taken, has to undress, must be carefully examined, and has to undergo all routine examinations. The physician must also remain hospitalized for a certain period after major surgery. Whenever the physician is seriously ill, he should not attempt to treat himself but let an experienced colleague take over instead.

The second remark pertains to the protection of physician's personal information in relation to the employer. Employees of public healthcare institutions normally have access to most or all healthcare data of patients that come for diagnostics or treatment. At the same time, we are bound to preserve confidentiality, and we are not allowed to share these data with others, for example with employers, without patient's consent. However, whenever a physician is being treated, his confidential

---

[1] http://www.itv.com/news/2016-10-01/world-medical-association-boxing-is-a-very-bad-idea/

information is available to other colleagues. If the director of a healthcare institution is a physician, he can view all healthcare data of fellow physicians on his computer. What advice to offer in situations of such conflicts of interest? Perhaps we should ask our IT experts to include in our database applications a transparent list of any persons who viewed the folder of a specific patient.

The final remark: we, physicians, are more vulnerable than we would like to admit. Even among us, alcoholism and addiction to other drugs (legal or otherwise) are not infrequent. We are doing ourselves a disservice if we are closing our eyes to the truth and if no one gathers enough courage and honesty to offer help to a colleague in distress. After all, who will take care of us if we do not help each other?

References

1. https://en.wikipedia.org/wiki/List_of_deaths_due_to_injuries_sustained_in_boxing
2. Ling H, Hardy J, Zetterberg H. Neurological consequences of traumatic brain injuries in sports. Mol Cell Neurosci. 2015;66:114–22. https://doi.org/10.1016/j.mcn.2015.03.012.
3. Purcell L, LeBlanc CM, American Academy of Pediatrics, Council on Sports Medicine and Fitness; Canadian Pediatric Society, Healthy Active Living and Sports Medicine Committee. Policy statement—boxing participation by children and adolescents. Pediatrics. 2011;128:617–23. https://doi.org/10.1542/peds.2011-1165.
4. White C. Mixed martial arts and boxing should be banned, says BMA. Br Med J. 2007;335:469.
5. Bosilj D, Cankar K. Physician as patient. Term paper for Dean's award. Faculty of Medicine, University of Maribor, 2017.

Further Reading

Almirall N. The ethics of engagement with the pharmaceutical industry. Mich Med. 2006;105(1): 10–2. PMID: 16555742.
Birden H, Glass N, Wilson I, Harrison M, Usherwood T, Nass D. Defining professionalism in medical education: a systematic review. Med Teach. 2014;36:47–61. https://doi.org/10.3109/0 142159X.2014.850154.
Braquehais MD, Tresidder A, DuPont RL. Service provision to physicians with mental health and addiction problems. Curr Opin Psychiatry. 2015;28:324–9. https://doi.org/10.1097/ YCO.0000000000000166.
Breitsameter C. How to justify a ban on doping? J Med Ethics. 2017;43:287–92. https://doi. org/10.1136/medethics-2015-103091.
Collins J. Professionalism and physician interactions with industry. J Am Coll Radiol. 2006;3: 325–32. https://doi.org/10.1016/j.jacr.2006.01.022.
Gould D. Gene doping: gene delivery for olympic victory. Br J Clin Pharmacol. 2013;76:292–8. https://doi.org/10.1111/bcp.12010.
Jawaid A, Rehman TU. Physician-pharmaceutical interaction: training the doctors of tomorrow. J Pak Med Assoc. 2007;57:380–1.
Kirschen MP, Tsou A, Nelson SB, Russell JA, Larriviere D, Ethics, Law, and Humanities Committee, a Joint Committee of the American Academy of Neurology, American Neurological Association, and Child Neurology Society. Legal and ethical implications in the evaluation and management of sports-related concussion. Neurology. 2014;83:352–8. https://doi.org/10.1212/ WNL.0000000000000613.
Leclerc S, Herrera CD. Sport medicine and the ethics of boxing. Br J Sports Med. 1999;33:426–9.

Marco CA, Moskop JC, Solomon RC, Geiderman JM, Larkin GL. Gifts to physicians from the pharmaceutical industry: an ethical analysis. Ann Emerg Med. 2006;48:513–21. Epub 2006 Feb 8. https://doi.org/10.1016/j.annemergmed.2005.12.013.

Mintzker Y, Braunack-Mayer A, Rogers W. General practice ethics: continuing medical education and the pharmaceutical industry. Aust Fam Physician. 2015;44:846–8.

Myers MF, Gabbard GO. The physician as patient: a clinical handbook for mental health professionals. Washington, DC: American Psychiatric Publishing; 2008. ISBN-13: 978-1585623129.

Negro M, Marzullo N, Caso F, Calanni L, D'Antona G. Opinion paper: scientific, philosophical and legal consideration of doping in sports. Eur J Appl Physiol. 2018;118:729–36. https://doi.org/10.1007/s00421-018-3821-3.

Reardon CL, Factor RM. Considerations in the use of stimulants in sport. Sports Med. 2016;46:611–7. https://doi.org/10.1007/s40279-015-0456-y.

Schetky DH. Conflicts of interest between physicians and the pharmaceutical industry and special interest groups. Child Adolesc Psychiatr Clin N Am. 2008;17:113–25, ix–x. https://doi.org/10.1016/j.chc.2007.07.007.

Trestman RL. Ethics, the law, and prisoners: protecting society, changing human behavior, and protecting human rights. J Bioeth Inq. 2014;11:311–8. https://doi.org/10.1007/s11673-014-9560-1.

Tucker AM. Conflicts of interest in sports medicine. Clin Sports Med. 2016;35:217–26. https://doi.org/10.1016/j.csm.2015.10.010.

Student Seminars

Abstract

Active student participation is the basis of teaching medicine, at both the undergraduate and postgraduate levels. This approach is especially important in teaching the course on medical ethics, in which only a minor part of the teaching consists of explanations of facts and data, and the major part rests on open discussion. In ethics, there are frequently no right or wrong answers. The emphasis is on judgment and justification, which must include every individual's values as well.

Seminar papers were prepared in the years 2009–2018 for the course *Medical Ethics and Law* at the Faculty of Medicine, University of Maribor. A team of five to eight students presented the seminars to their colleagues and the teacher, followed by an open discussion. The papers are listed in the order they were presented to students.

Most seminars deal with ethical problems that are of importance for a student of any nationality. Seminars on topics of local interest are also included. While they may not be directly applicable to the teaching of medical ethics in other countries, the reader will get an idea about the broad field of topics that students of medicine should be able to discuss.

Cases were often prepared on the basis of real events, known to the author during his professional work or as a member of various ethical committees. The identity of the persons involved has been masked, except for cases that were widely discussed in the media.

Teaching medical ethics is a rewarding activity. On the photo: Dr. Matjaž Zwitter with the class 2017/2018, Medical Faculty, University of Maribor

List of Students' Seminars

1. Surrogate motherhood
2. Physician's confidentiality
3. Eluana Englaro
4. Love life
5. Donor for bone marrow transplantation
6. Unsolicited medical intervention
7. Elderly driver
8. Posthumous insemination
9. Physician as patient
10. Crime as mental disease
11. Conscientious objection
12. Ethics and gladiators in professional sports
13. Medical malpractice and the right to compensation
14. Anorexia nervosa
15. Prevention of pregnancy in psychiatric patient
16. Placebo in clinical trials
17. Choosing the gender of the newborn
18. Vegan diet for children
19. Accusation of medical malpractice—disclosure of personal information
20. Boxing
21. Gifts
22. Drug addicts, pregnancy, and parenthood
23. What makes a good physician?
24. Parents declining mandatory vaccination of their children
25. Collaboration between the psychiatrist and the family physician

26. Medically assisted insemination for healthy women
27. Physicians as leading politicians
28. Shooting as an Olympic sport
29. Fine-needle biopsy of the breast for a 12-year-old girl
30. The death of Ivan Ilyich
31. Waiting periods for funerals
32. Cancer ward
33. Physician-alcoholic
34. Disappearance of inexpensive drugs with long-lasting positive experience
35. Empathy and trust
36. Medical treatment of patients without healthcare insurance
37. Doping in sports
38. Intimate relationships with a patient
39. Paulo Coelho: Veronica decides to die
40. Communication with a troublesome patient
41. The Franja Partisan Hospital
42. The donor of embryonic cells and anonymity
43. Addiction with prescription drugs
44. Literature as a medication
45. Sinclair Lewis: Arrowsmith
46. Female circumcision
47. Accusation of medical malpractice
48. Transport of the dying, chronic patient to the emergency department
49. Dr. Catherine Hamlin
50. An aggressive patient
51. Legalization of marihuana
52. Lay people's attitudes towards euthanasia
53. Discrimination
54. Is pedophilia a disease?
55. The Nuremberg trials against the Nazi physicians
56. Humor in communication with patients
57. Obamacare—American healthcare reform: Successes and difficulties
58. Child abuse
59. Ethical questions in self-inflicted diseases
60. Homeopathy
61. Treatment of the demented patient
62. The physician in commercials
63. The Ebola epidemic—ethical questions
64. Gene testing in underage daughters
65. Revocation of driver's license
66. Airplane seats for overweight persons
67. Individual consent for review of old biopsies
68. Love in a nursing home
69. Professional sports in children
70. Medical strike

71. Artificial womb
72. Genetic testing for prediction of a disease
73. Cancerophobia
74. Communication in the waiting room
75. Traditional medicine
76. Late termination of pregnancy
77. Postponement of prison sentence service for health reasons
78. Animals in biomedical research
79. Death of grandparents
80. Guerilla surgeon
81. Loneliness
82. Mental health of political leaders
83. Eugenics
84. Fatherhood
85. Mark Langervijk
86. Molière and the characters of physicians
87. Alternative diagnostics
88. Trade with human organs for transplantation
89. Dr. Thomas Percival: Medical Ethics
90. F. Mielke, A. Mitscherlich: Doctors of Infamy—The Story of the Nazi Medical Crimes
91. Oregon: The Death with Dignity Act

22.1   Surrogate Motherhood

Helena, 27, has been married for a year and has no children. After a diagnosis of cervical cancer, stage IB, the proposed treatment is radical radiotherapy and concurrent chemotherapy. According to her oncologists, prospects for long-term survival are relatively favorable, and the expected 5-year survival rate with no evidence of recurrent disease is 80%.

Helena would like to have a baby, but it is certain that she will be infertile after treatment. Helena wants her egg cells removed prior to treatment and have them inseminated with the semen of her husband. Helena's mother Barbara is 45 years old, healthy, and still has regular periods. Barbara agrees to be the surrogate mother for Helena's baby. An inseminated egg cell would be implanted in Barbara's womb; she would give birth to the child and then give the child to her daughter.

Task
Present the case and focus on the following questions:

- What are the legal and ethical provisions of surrogate motherhood?
- Can we accept the suggestion for surrogate motherhood in this concrete example?

- How would our opinion change if Helena's sister (age 26, unmarried) were to be the surrogate mother? Alternatively, Helena's friend who is 38 years old and already has two children of her own? Alternatively, an unknown healthy woman who would receive financial compensation for surrogate motherhood?

22.2 Physician's Confidentiality

- 1992: Gorazd and Veronika married and chose to have the same family physician.
- 1998: The couple divorced. Veronika moved away, but she kept the same family physician.
- 2000: Gorazd told his physician that he has been leading "a very free life" since the divorce. Blood tests confirmed that he is HIV positive. Gorazd received appropriate treatment and recommendations.
- 2001: During a control check-up, Gorazd let the physician know that he had reconciled with Veronika and that his ex-wife was moving back in with him. The physician made his best effort to convince Gorazd to tell Veronika about his infection or to ensure that she could not be infected. Gorazd ignored these recommendations and left the physician's office without giving a guarantee that he would follow the recommendations.

Task
Present the case. Questions:

- Should the physician disclose information on her ex-husband's infection to Veronica? Alternatively, should the physician respect medical confidentiality?
- Discuss with a physician specializing in infectious disease the issues of confidentiality in patients with sexually transmitted diseases. Present the ethical benefits/burdens in the physician's decision.

22.3 Eluana Englaro

After an injury in a traffic accident, Eluana Englaro remained in a vegetative state for 17 years. At the request of her parents and in line with the court decision, supportive care with tube feeding and supply of liquids was discontinued. After 1 week, on February 9, 2009, she passed away.

Task
- Briefly present the case.
- Present the arguments in favor of continuing medical support—feeding and supply of liquids.
- Present the arguments for discontinuing supportive care.
- Is there a difference between termination of feeding and supply of liquids and euthanasia?

22.4   Love Life

Ray Kluun, Macmillan 2007.
  One of the most shocking examples of euthanasia.

Task
Present the book and present the story of Carmen: disease–treatment–dying–death.
Discussion:

* Decision for euthanasia: is this really an act in respect of patient's autonomy?
  Would it be different, in terms of autonomy, if we let Carmen die without a physician's intervention?
* To what degree is her decision for euthanasia influenced by others: husband,
  relatives and friends, physician?

22.5   Donor for Bone Marrow Transplantation

A 5-year-old Adrian has been diagnosed with thalassemia. Bone marrow transplantation is by far the best treatment option. To avoid transplant rejection, the donor and
the receiver must be histocompatible.
  In the past, bone marrow was taken from pelvic bones under general anesthesia.
Nowadays, immature hematopoetic stem cells are collected from peripheral blood,
a procedure which is considerably less burdensome for the donor. Hematopoietic
stem cells for bone marrow transplantation can also be obtained from the placenta
immediately after the birth of a child. In this case, the donor (the newborn) is not
harmed in any way.
  Possible bone marrow donors are:

* One of the parents
* Adrian's 12-year-old brother
* Adrian's 2-year-old sister
* Adrian's future brother or sister

  Statistically speaking, there is a 25% chance that the future child of the same
parents will be histocompatible with his or her already born brother or sister. This
probability can be considerably increased, however, if we perform in vitro fertilization and then implant in the womb an embryo that is histocompatible. At the same
time, in vitro fertilization allows us to guarantee that the future child will not develop
thalassemia. In this case, no intervention in the newborn child is required because
bone marrow stem cells are obtained from the umbilical blood and the placenta.

Task
Present the case. Discussion:

* What ethical burdens can we predict for each of the possible bone marrow donors
  listed above?

- Where should we draw the line between ethically acceptable and ethically unacceptable tissue donors for transplantation?
- Let us assume that the first three options are ruled out because of incompatibility. If so, is it ethically acceptable to propose that the parents have another child chosen so that he or she can, in addition, provide "spare parts" for Adrian?

22.6   Unsolicited Medical Intervention

A physician is going home on a public bus. Next to her stands a middle-aged woman. The physician notices that at the nape of the woman's neck, there is a black and brown nevus. Based on its visual aspects, the physician strongly suspects this is a malignant melanoma. Should the physician warn the woman, whom she does not know, about her mole?

Task
Present the case. Discussion:

- Ethical analysis—ethical burdens and benefits of the intervention.

See also M. Zwitter et al. Professional and public attitudes towards unsolicited medical intervention. Br Med J 1999; 318: 251–253.

22.7   Elderly Driver

Anton, a retired train conductor, is 72 years old and lives together with his wife and the family of their 43-year-old daughter. Over the past year, family members have noticed that Anton is becoming increasingly uncertain in daily household chores and occasionally gets lost. On a few occasions, he could not find his way back home. Medical examination revealed high blood pressure and atherosclerotic occlusive disease. A brain CT scan showed numerous subcortical perivascular lesions. Anton was diagnosed with moderate subcortical vascular dementia.

Despite the physician's explanations, Anton denies any mental troubles and firmly rejects advice to quit driving. According to him, the recent minor damage to the vehicle occurred due to the vehicle's age. Anton surprises his family by telling them that he has just decided to buy a new car.

Should the physician break medical confidentiality and report Anton's inability to drive safely to the authorities?

Task
- Visit three family physicians and ask them what their course of action in such situations would be. Have they encountered such a case in their practice?
- Explain why it is important to distinguish between the role of the family physician and the role of a physician as an expert (e.g., in issuing a medical certificate for a driver's license).

22.8   Posthumous Insemination

In 2008, when Branko divorced and moved in with his new girlfriend, his son was 12 years old.

Five years later, Branko was diagnosed with acute leukemia. Before starting the treatment, he and his girlfriend decided to freeze his semen, so that they could have children after treatment. One year after intensive treatment, leukemia recurred. The second-line treatment resulted only in short-term remission and physicians informed Branko that there were no chances for lasting improvement.

Upon his last visit to the hospital, Branko took the physician and the nurse by surprise and asked them to sign his will as witnesses. Because Branko was fully able to reason clearly, the physician agreed to co-sign the will. A few weeks later, Branko passed away.

After Branko's death, they read the will: Branko left all his possessions to his ex-wife and his son.

Branko's girlfriend asked to be inseminated with his frozen semen.

Task
Present the case. Discussion:

- Was it appropriate for the physician and the nurse to sign the will?
- Should the girlfriend's request to be inseminated with the frozen semen be approved?

22.9   Physician as Patient

Even physicians get sick. We often do not apply the same rules to ourselves as we do to patients. The purpose of this seminar is to highlight these differences from the medical and ethical perspectives.

Task
Visit the administration of the university hospital and ask for the data on average sick leave in the past year for physicians, for other medical staff, and for non-medical staff.

- Are physicians stricter when it comes to approving sick leave for physicians as opposed to other patients?
- Are physicians given priority in medical treatment? Ask five physicians working in diagnostic branches of medicine (radiology, nuclear medicine) and five working in therapeutic branches (internist, surgeon) if they would consider bypassing the waiting list to admit a colleague physician. If the answer is "yes," is such privilege acceptable from the ethical point of view?
- While standards for diagnostic and therapeutic procedures are normally respected for all other patients, there is anecdotal evidence that physicians sometimes

ignore these standards when treating their colleagues. Are physicians equally, more, or even less careful in communication, diagnostics, and treatment of colleagues, as compared to other patients? Pose this question to the above-mentioned physicians.

22.10 Crime Due to Mental Disease

In the case of Josef Fritzl, who was accused and convicted of murder, rape, incest, and enslaving his daughter and their children, an expert psychiatrist concluded that Fritzl suffered from a severe combined personality and sexual disorder. Fritzl was sentenced to life imprisonment in a prison for psychiatric patients. A detailed description of the case can be found on Wikipedia.

According to the opinion of some psychiatrists who examined him, Norwegian mass murderer Anders Behring Breivik is a mental patient.

Task
Present the case. Questions:

- If we follow this line of reasoning, any criminal can be diagnosed as a mental patient. Is this appropriate?
- What are the implications for psychiatric patients and their status in society when a serious criminal is publicly diagnosed with a psychiatric diagnosis?

22.11 Conscientious Objection

Prior to applying for specialization in gynecology, a physician had declared conscientious objection. She rejects any medical intervention in human reproduction. She, therefore, rejects contraception, abortion (with the exception of cases when the mother's condition is directly threatening her life), prenatal diagnostics, and medical assistance in infertility.

Currently, this physician works as a gynecologist in primary care. Women who choose her as their gynecologist are informed in advance which procedures the physician refuses to perform or does not give advice on. They are also informed about those procedures for which the physician will not issue a referral to another gynecologist with the exception of life-threatening conditions.

Task
Present the case. Questions:

- Given the openly expressed conscientious objection, would it be appropriate if the committee of the Medical Chamber of Slovenia had not approved her specialization in gynecology?

- How does the physician's conscientious objection affect relationships with her colleagues employed in the same institution?
- Is it appropriate that the physician refuse to give women advice regarding medical assistance in cases of infertility?

22.12   Ethics and Gladiators in Professional Sports

Everyone in Slovenia knows the story of cross-country skier Petra Majdič: despite being injured with broken ribs, she won a bronze medal at the Olympics in Vancouver. During the warm-up before the race, Petra skid off-course, fell into a deep gulley and suffered a severe blow to her chest. Medical examination immediately after the accident did not show any serious injury. Despite excruciating pain, she took part in the competition. Sports fans praise her as a role model who had an incredible will to overcome pain and win a medal. However, people rarely consider that the course of events might have been different. After the race, it turned out that Petra had four broken ribs and a pneumothorax. If her spleen had been torn, we could have seen a live broadcast of a life-threatening emergency.

Task
Present the story of Petra Majdič and her injury. Questions:

- What are the limits of risk-taking in professional sports?
- Discuss the responsibility of a physician who is entrusted with the care of professional athletes.

22.13   Medical Malpractice and the Right to Compensation

Mateja is 27 years old. On Saturday night, she went to the hospital due to severe stomach pain, vomiting, and increased temperature. She was admitted to the department of surgery and was diagnosed with appendicitis. A specialist registrar on duty who had been continuously working for 21 h informed a senior specialist registrar on duty and proceeded with Mateja's surgery. An appendectomy was performed without complications. Mateja left the hospital 4 days after surgery. One week after surgery, however, nausea, stomach cramps, and occasional diarrhea recurred. The surgeon's examination and radiological diagnostics did not give a clear picture of her problems. After 8 weeks, surgeons, therefore, decided on another surgery. During the surgery, they found the cause of Mateja's issues right between the folds of the intestines: a piece of surgical material that had remained there after the first surgery.

There is no doubt that Mateja suffered damage to her health due to the insufficiently careful work of the physician; she was operated on twice and had to stay on sick leave for 3 months. At the same time, we know that there is no physician who has never overlooked something or made a wrong decision.

Task
Present the case. Questions:

- Is it fair to blame the young surgeon who has been working continuously without rest for 21 h for the malpractice?
- Does the patient have the right to compensation for injury incurred during the treatment in cases of rare, unexpected complications and injury after the surgical procedure?
- Does the patient have the right to compensation even if the physician is not proven guilty (no-fault compensation)?
- If the right to compensation depends on proving the physician's guilt: what does this imply in terms of honesty and communication between the physician and the patient? Are we not in doing so forcing physicians to conceal their mistakes and to shy away from analyzing and ameliorating them?

22.14 Anorexia Nervosa

Andreja had been struggling with anorexia since high school. During her medical studies, she passed all her exams with excellent grades, yet, her progress was very slow. At the age of 28, when most of her classmates had already finished their studies, she was still in her 4th year. After two hospitalizations due to severe exhaustion, her condition severely deteriorated for the third time. At first, she firmly rejected another hospitalization. After several hours of persuasion, she finally promised her friends that they could take her to the hospital the next morning. It was too late. She died during the night.

Task
Present the case. Questions:

- Is force-feeding of patients with anorexia nervosa allowed and, if so, under what conditions?

See also: Hébert, P. C., Weingarten, M. A. The ethics of force-feeding in anorexia nervosa. Can Med Ass J 1991; 144: 141–144. http://www.ncbi.nlm.nih.gov/pmc/articles/PMC1452985/.

22.15 Prevention of Pregnancy in Psychiatric Patient

Anica is 34 years old and has been treated for paranoid schizophrenia for 12 years. She is divorced and has no children. She lives together with her mother (66 years) and her father (72 years). After the diagnosis, she was first treated with neuroleptics in the form of pills. Anica interrupted her treatment several times; subsequently, her mental state deteriorated and she had to be hospitalized again. For the last 4 years, she has been receiving therapy in the form of depot injection once a month.

For the last 2 years, Anica has had a friend whom she met at the out-patient department of the psychiatric hospital. Her parents were delighted as the burden of an adult, but not independent, daughter was too heavy for them. During this period, Anica had two miscarriages. She rejects any discussion of contraception and wants to have a baby. However, her mother knows that a grandchild would only add to the overall burden, which is why she asks the gynecologist and psychiatrist to give her daughter, during her regular therapy at the psychiatry department, a depot injection of progesterone to prevent pregnancy.

Task
Present the ethical analysis of the three possible scenarios:

- Anica becomes pregnant again and gives birth to a child whom she is not capable of taking care of. Potential adoption would be difficult due to a genetic predisposition for schizophrenia.
- Along with neuroleptics, Anica receives a depot injection for hormonal contraception without consenting to it.
- They prevent Anica from seeing her friend in the future.

22.16 Placebos in Clinical Trials

Placebos play an essential role in ensuring the objectivity of clinical trials. Nevertheless, we have to realize that people lacking medical education have a poor understanding of what a placebo is; this is especially true for persons with limited autonomy.

As a member of a local ethical committee, you are evaluating a proposal for a clinical trial that involves psychiatric patients with a bipolar disorder. In a randomized trial, a new drug would be compared to a placebo.

Task
Present the scientific basis for the use of placebos in randomized clinical trials. Questions:

- When is the use of placebos in a clinical trial ethically acceptable?
- Discuss the ethical dilemmas of using placebos for psychiatric patients.

22.17 Choosing the Gender of the Newborn

In many countries, especially in Asia, there is a widespread conviction of the superiority of males over females, due to which many parents wish to have a male child. Such distorted views on the superiority of the male gender lead to severe abuses: selective abortions of female fetuses, feticides, and murders of newborn girls.

Task

Present global figures on male/female ratio at birth and discuss statistics or estimates regarding the number of female feticides and infanticides. Questions:

- Is it ethical to disclose the gender when the physician can legitimately assume that an abortion will be performed in the case of a female embryo?
- Is it allowed to perform gender-based embryo selection during in vitro fertilization? Under what conditions would you allow it? Only in cases of the gender-dependent genetic disease, at the parents' demand, or never?

22.18   Vegan Diet for Children

Every adult mentally competent person has the right to choose his diet, regardless of how problematic it may be. While proponents of vegan diets emphasize some selected positive aspects of their choice, they neglect the long-term consequences of severely unbalanced diets, especially for children.

Task

Explain what a vegan diet is and compare it to the essential nutritional requirements for humans. Questions:

- What are the risks of a vegan diet for a child?
- Present the official position of the leading international associations of pediatricians regarding vegan diet for children.
- Do the parents have the right to pass their choice of unbalanced and potentially harmful diet to their children?
- Is society obliged to offer a vegan diet at the parents' request in public institutions (nursery schools, schools, hospitals)?
- What should a pediatrician do to protect a child of vegan parents?

22.19   Accusation of Medical Malpractice: Disclosure of Personal Information

In accusations of medical malpractice, it often happens that the victims publicly share their views of the medical dispute. They reveal the physician's identity even before any discussion in the professional community takes place and prior to potential court proceedings.

Task

Present the physician's commitment to confidentiality and discuss the situations in which confidentiality may be dismissed. Questions:

- Is the physician allowed to defend him or herself against public accusations also by disclosing those parts of the patient's or family's personal information that are crucial for objective judgment but would normally be considered a confidential?
- Let us consider a concrete example: is the physician allowed to reveal in public the patient's drug addiction, the fact that it was not indicated to the physician by the patient's relatives upon admission, and that this factor crucially contributed to treatment complications?

22.20 Boxing

Available data prove that boxing causes not only acute injuries but also chronic brain injury. This applies not only to boxing as a professional sports activity but also to boxing among amateur athletes since the helmets they use cannot protect against counter-punch injuries of the brain hitting the skull.

Task
Present typical acute and chronic injuries linked to boxing. Questions:

- Is it ethical for a physician to be actively involved in boxing (medical examination before the career, medical examination and advice during boxer's career, physician's presence in fights)?
- Present the position of the British Medical Association towards boxing. You are strongly encouraged to also present the viewpoints of Slovene physicians (sports medicine specialists, neurologists). If your group does not reach an agreement about a particular question, present the arguments for one or the other viewpoint.

22.21 Gifts

Occasional gifts to physicians and other healthcare staff are a relatively frequent phenomenon.

Task
Visit five to ten physicians in general practice and at the university hospital and ask them about gifts that they recently received, assuring them confidentiality. Questions:

- What are the boundaries between attentiveness, gratitude, and bribery?
- When are we allowed to accept a gift and when should we decline it?
- Who is frequently receiving gifts, who rarely or never does?
- How can the acceptance of gifts disturb the work of a medical team?

22.22   Drug Addicts, Pregnancy, and Parenthood

Present the problem of drug addicts, especially those on methadone therapy. Get in touch with a center for the treatment of drug addiction. We are especially interested in ethical questions regarding pregnancy and parenthood.

Task
Visit your nearest Drug Addiction Treatment Center and ask the personnel about drug addicts who became parents. Questions:

- Is parenthood helping drug addicts to find their purpose of life?
- Are they capable of taking care of the child?

22.23   Who Is a Good Physician?

What are the qualities that make a good physician? Answers will vary considerably depending on whom we ask: other colleagues-physicians, head of the department, nurses, patients. We are interested in learning how people outside the medical profession imagine a good physician.

Task
The first group will survey 50 healthy persons. Recruit healthy individuals older than 30 years of age; you can invite your colleagues, friends, parents, neighbors, or passers-by. First, ask the interviewees to name three to five qualities that they consider crucial for someone to be considered a good physician. You will then ask the interviewees to circle on a provided list of options the three most important qualities of a good physician. Compose the list according to your own judgment (e.g., meticulous, precise, on time, always in a good mood, leading expert, modest, kind, talks clearly, reliable, compassionate). Note the gender and age of every interviewee.

The second group will carry out the same survey among patients. Recruit approximately 50 participants in a waiting room of the healthcare institute or hospital. Announce your visit in advance to the head of the center or the head of the nursing staff.

Present the survey results and compare the results you obtained for healthy persons and for patients.

22.24   Parents Declining Mandatory Vaccination of Their Children

Present the two views on mandatory vaccination: public interest to limit or eradicate certain contagious diseases and protect those children who should not be vaccinated for medical reasons, and the position of vaccination opponents who emphasize the dangers of vaccination.

Compared to the issue of a vegan diet in children where only an individual child and his or her best interest was considered, here, the interest of society as a whole is at the forefront.

Task
Present arguments for and against mandatory vaccination. Questions:

- To what extent it is legitimate to put the interest of society before the interest of the individual?
- Present the question from the point of view of public health.
- Can the negative standpoint of the vaccination opponents have positive consequences?

22.25 Collaboration Between the Psychiatrist and the Family Physician

For examination or treatment with a psychiatrist, the patient does not need a referral from the family physician. Psychiatrists send the report on examination to the patient, but not to the patient's family physician. It is, therefore, the responsibility of the patient to either communicate the information about his diagnosis and treatment with the psychiatrist to his physician or to keep the information for himself.

Task
Visit three family physicians and ask them about their standpoint towards such limits to the availability of professional information. Questions:

- Is it appropriate if the personal/family physician is not aware of all the medications the patient is receiving?
- Who should be held responsible in cases of drug interactions?
- Discuss the responsibility of a family physician or of a public health specialist for issuing a medical certificate, as needed for a driver's license or gun permit.

22.26 Medically Assisted Insemination for Healthy Women

In Slovenia, the right to medically assisted insemination for healthy women was subject to a nation-wide referendum. At that time, the majority voted against allowing such a procedure.

Task
Briefly present arguments for and against medically assisted insemination, using published opinions prior to the referendum. Questions:

- Is having a child a right that should be guaranteed to everyone by the state?
- Is non-parenthood in adult persons a state of disease that should be addressed by the healthcare system?

- Leaving financial questions aside, is there any ethical problem regarding medically assisted insemination for a healthy woman?
- If we grant the aforementioned right to homosexual women, are we not discriminating against two homosexual men who might also want to have a child?
- A logical follow-up to the previous question: to prevent discrimination against male homosexual couples, we would have to support studies on embryo development in artificial wombs. In that way, men could also have children conceived in vitro from their semen and a donated egg cell.

22.27   Physicians as Leading Politicians

It is not rare for a physician to become a leading politician:

- Dr. Salvador Allende (Chile)
- Dr. Radovan Karadžić (Republika Srbska)
- Dr. Ernesto Che Guevara (Cuba)
- Dr. Bashar al-Assad (Syria)
- Dr. Sali Berisha (Albania)

Task
Present a few stories of physicians as political leaders. Invite your colleagues from one of the social sciences faculties (sociology, political science) to your presentation. They might help us in the discussion.
Questions:

- Can you think of any other physician who became a statesman?
- Can we find common traits among the physicians who moved up to become leading politicians?
- Are physicians more, or perhaps less humane once politics grants them the power to decide about the lives of masses of people?

22.28   Shooting as an Olympic Sport

Sports and especially amateur sports should support people's physical and psychological health. If this is how we understand the mission of sports, we may wonder if it is appropriate that shooting sports is (still) an Olympic sport.

Firearms are used for two purposes: for killing wild animals and for killing people. Both of these purposes are ethically very controversial. Almost no one nowadays needs to rely on hunting for food supply or as a defense in the wilderness. Quite the opposite: given that the terrain on which animals can live freely is becoming ever scarcer, hunting is more and more reflecting a sadistic and arrogant attitude towards the natural environment. A great majority of firearms, however, is made for killing people. From the ethical standpoint, we cannot agree with a sport that improves the ability to kill people.

Task
Present the arguments for and against shooting as an Olympic discipline. Present the goals of the Olympic movement and determine whether the shooting is in line with these goals.

22.29 Fine-Needle Biopsy of the Breast for a 12-Year-Old Girl

A 12-year-old girl came for a fine-needle biopsy of the breast. She was referred by her general practitioner and accompanied by her mother. The cytopathologist was convinced that this was just a case of asymmetric early-puberty breast swelling. He explained to the mother that the girl was fine and that, in his opinion, a puncture was unnecessary, potentially harmful. Fearing that her daughter has cancer, the mother insisted on having a fine-needle biopsy of the breast performed.

Task
Visit the department of cytopathology at the university hospital and ask the physicians about their experience on questionable indications for biopsy. Questions:

- Should the cytopathologist perform breast puncture even though he was convinced that it was unnecessary and potentially harmful?
- Should physicians working in diagnostic services strictly follow the indication, as defined by a referring physician, or should they decline to perform an unnecessary procedure?

22.30 The Death of Ivan Ilyich

Leo Tolstoy wrote a story titled "The Death of Ivan Ilyich."

Task
Read the story. Present the character of Ivan Ilyich, his slowly progressing disease, and death. Describe Ivan's relationship with his family, colleagues at work, and physicians before the disease and during times of his progressively declining condition. Can we find similar examples of patient distress and relationships in the present days as well?

22.31 Waiting Periods for Funerals

In Stockholm, the relatives of the deceased often have to wait even more than a month for the funeral to take place.

Task
Find the data on funeral waiting periods. Contact medical students from Stockholm and ask them to share first-hand their account of how they view the problem.

- Are funeral waiting periods an ethical question?
- What burden do they represent for the family and friends of the deceased?

22.32   Cancer Ward

The book "Cancer Ward" by Alexander Solzhenitsyn is one of the most powerful accounts of all the layers of relationships in a hospital.

Task
Read the book. Present the relationships among the patients and the relationships between patients, nurses, and physicians. These relationships are dynamic rather than being static; they are also influenced by the development of a disease.

Select a few typical persons from the book and present them to your colleagues.

22.33   Physician-Alcoholic

There can be no doubt: physicians are not immune to alcohol abuse. Alcohol abuse is not only the problem of an individual physician or his family; it affects the entire staff of the healthcare institution and, of course, the patients. A relatively easy task is to act against a colleague who would be drunk when coming to work, or who would drink alcohol while being in service. However, we should also think of colleagues who regularly drink in their free time, in which case their physical and mental capabilities will inevitably deteriorate.

Task
Try to explain the paradox: alcohol is among the most popular gifts given to physicians, yet no patient would tolerate a physician who is an alcoholic. Further questions:

- Where is the boundary between solidarity among physicians and unacceptable tolerance of a colleague's alcohol addiction?
- Are we helping him if we are closing our eyes to the issue as colleagues?
- How can we approach a colleague who fiercely denies his addiction?

22.34   Disappearance of Inexpensive Drugs with Long-Lasting Positive Experience

Along with the common experience of new drugs, one rarely hears that the old, quality drugs, proven over the course of decades, are being left to oblivion. Many pharmaceutical companies are giving priority to the new, even up to ten times more expensive (and frequently not much better) drugs, while the old drugs disappear from the market.

Task

Ask the head of the pharmacy in the university hospital to provide you with a list of medications that were removed from the market by their manufacturers. Questions:

- Ask some of the *older* internists whether the new drugs are really that much better than the older ones. Can they name a drug that is no longer available on the market but was considered good, safe, and trusted?
- Ethical question: is it appropriate for the pharmaceutical company to remove a drug from the market due to its low price even though physicians would still prescribe it?

22.35 Empathy and Trust

Empathy from the physician's side and trust from the patient's side are essential for a good relationship.

Task

Try to explain what empathy is. During the practical courses of medicine, you have certainly established stronger ties with a patient that became dear to your heart and whom you remember much deeply and not just as a case of a disease. Questions:

- Is this the positive side or the burden of our profession?
- How is the physician's empathy different from the compassion that we as human beings feel towards each other?
- How does the physician's empathy relate to the patient's trust?
- What are the potential obstacles to empathy and trust?

22.36 Medical Treatment of Patients Without Health Insurance

On May 29, 2010, a 48-year-old worker from Macedonia had severe chest pain and visited the emergency department. Because he had no health insurance, he was informed that he would have to cover the costs of examination himself if it turned out after examination that it was not an emergency situation. He went back home. After a couple of hours, the man died due to myocardial infarction.

Task

Present the case, as presented in the media. Questions:

- What criteria can we use to distinguish between urgent and non-urgent medical treatment?
- May we assign the label "unnecessary and non-urgent" to a patient with chest pain for whom medical examination does not confirm a serious disease?

- Before examining the patient, how can the medical staff know whether the case is urgent?
- Who is responsible for the unfortunate case presented above—the staff of the healthcare institution that followed the rules or those who prepared the rules?

22.37 Doping in Sports

As many sports have evolved into a big business, spectacular results are more important than the health of the athletes.

Task
Present an overview of doping in competitive and in recreational sports. Questions:

- Where to set the boundary between legal and illegal stimulants?
- Are the sportsmen the only one to blame (e.g., Lance Armstrong), or should we also blame those who remain in the background but gain huge profits from spectacular achievements?
- What is the responsibility of sports physicians?
- What about financial interest and the responsibility of the pharmaceutical industry?
- Would it make sense to require that certain drugs (e.g., erythropoietin) remain traceable all the way from the manufacturer to the end user (patient)—as is done for narcotics—and to thus limit the possibility of their abuse in sports?

22.38 Intimate Relationships with a Patient

Miloš is a 34-year-old specialist in family medicine in a larger Slovenian town. He is married and has three children, aged 7, 5, and 1 year.

For about 2 years, one of his patients has been 23-year-old Barbara. She grew up in unsettled family circumstances in which alcohol and fights were frequent. She managed to finish a vocational secondary school for catering and move away from her parents. She remains very insecure, works on a fixed-term contract, and with irregular salary. Since she split with her boyfriend because she caught him with drugs, she has no real friends.

Initially, she visited her physician due to depression. Later she made several visits due to insomnia, digestive disorders, and lower back pain. The physician attempted to comfort her by saying that everything in life will work out eventually. The first time he gave her a friendly hug, which later became a regular part of her visits. The past fall, she called him telling she was at home feeling very ill. The physician offered to come and visit her at her home. The visit concluded with sexual intercourse, to which Barbara consented freely. In the next months, the physician visited her at least once a week.

Barbara asked him to break off their relationship and expressed the wish to choose a new personal physician. The physician did not want to hear about it and insisted that they keep on seeing each other.

Task
Present the story and answer the following questions:

- Was the physician's behavior against the code of medical ethics?
- Was his behavior against the law?
- What principled viewpoint and recommendation can we adopt when it comes to intimate relationships between medical staff and patients?

22.39   Paulo Coelho: Veronica Decides to Die

The novel by Brazilian author Paulo Coelho is set in Slovenia: it is a story about Veronica who tries to commit suicide. When she wakes up after a few days, a psychiatrist tells her that the sleeping pills had caused permanent damage to her heart and that she had at most 1 week to live. The nearness of death changes Veronica's attitude towards life.

Task
Read the book. Present its contents and your reflection on it.

22.40   Communication with a Troublesome Patient

In their attitude towards physicians and other personnel, patients and their relatives reveal a whole spectrum of attitudes. While most patients expect decent service within accepted rules, some may be described as annoying, unpleasant, and troublesome. Such patients typically see only themselves and do not understand that physicians must distribute their work and time across all their patients; they often come to the office without making an appointment, or they show up at times when the physician is finishing the work for the day; they complain over trivialities; they cannot accept the fact that the physician must follow the agreed-upon rules when making decisions on diagnostics, treatment, and sick leave; they sometimes record the conversation or even visit the physician together with their lawyer.

Task
Visit your personal physician and ask him/her to tell you about such unpleasant patients. Present the cases in an anonymized form. Make sure you change the data sufficiently so that the patient's identity remains reliably undisclosed. Questions:

- What distinguishes an unpleasant ("troublesome") patient from a patient who is simply very worried about his or her own health?

- How should the physician manage communication with such a patient?
- To what extent should the physician give in to the patient's demands?
- How should the physician protect himself and his staff against potential subsequent accusations? Emphasize the role of the medical team and of the head of the department in the prevention and resolution of conflicts.

22.41   The Franja Partisan Hospital

In Nazi-occupied Europe, the Slovene Partisan healthcare represented a unique example of medical work undertaken in extremely modest conditions. It was the epitome of solidarity, courage, and ingenuity. The Franja Partisan Hospital is a monument of Partisan healthcare and is preserved as a UNESCO European cultural heritage monument.

Task

Read up on Partisan healthcare. Visit the Franja Hospital and prepare a presentation. Consult Milojka Magajne, a professor of history and sociology at the museum in Cerkno. She will be happy to help you. According to the latest data, only one of the nurses who worked at the Franja Hospital and a small number of patients who were treated there are still alive. Milojka Magajne will be able to tell you if you can visit the nurse (she is 92 years old) or one of the patients. If you manage to arrange an interview with one of the persons mentioned above, that would be a very interesting first-hand account for all of us.

22.42   Donor of Embryonic Stem Cells and Anonymity

A summary of a story that will be published in *The Newcastle Chronicle* on August 15, 2028.

Andrew is a 43-year-old construction engineer, manager of *Quick Constructions Ltd.*, married and a father of three children. He received a letter in which 22-year-old Michael asks him for a meeting. At the meeting, Michael tells him that he, Andrew, is his biological father. Andrew only has vague recollections that in their college days he and three of his colleagues indeed donated their semen for insemination of women whose partners were infertile. He remembers that this was great fun to them, but he does not recall whether he signed any agreement allowing disclosure of his identity to the future child upon reaching the age of majority. Michael expects Andrew to admit his fatherhood, which could, however, have consequences for relationships in Andrew's family.

In 2005, Great Britain passed a law giving the child who was conceived with donated semen by means of medically assisted insemination the right to find his or her biological father once he or she reaches the age of majority. In many countries, semen donors remain anonymous. You can read more about the procedures for sperm and egg cell donation on the web.

Task
Present the legislation that regulates the issue of embryonic cell donation and lays out the conditions for donor anonymity or his/her identity disclosure. Questions:

- What are the legal consequences of disclosing the donor's identity (biological father or mother)? Who can benefit from this procedure if anyone?
- What are the possible abuses?
- Is the potential agreement to donor identity disclosure still valid after 20 years?

22.43 Addiction to Prescription Drugs

Monika, 43, works as a nurse at one of the surgery departments of the university medical center. She has two daughters, both students. Five years ago, she divorced since her husband had not quit alcohol and had continued showing aggressive behavior towards her and their children. Her employees describe her as diligent, somewhat introverted, but otherwise having good relationships with her colleagues.

Half a year ago, the clinic decided to stop procuring large stocks of medications in departments. Instead, they transitioned to the continuous daily supply of medications from the central pharmacy. Along with this change, it became evident that this clinical department uses unusually large amounts of the drug alprazolam. An internal investigation was carried out which revealed that the drug disappeared whenever Monika worked the night shift. At first, she denied all charges but eventually admitted the abuse. She explained that she was first given the drug by her personal physician when she was in distress. Along with the medication that she took from the departmental pharmacy, she regularly obtained prescriptions from her personal physician. In addition, she frequently asked the physicians from her department for prescriptions with the excuse that she needed the medication for her relative. She is currently taking at least ten pills per day.

Task
Present the problem of addiction to prescription drugs, especially in relation to other forms of addiction (alcohol, illegal drugs) which are more frequently discussed. Questions:

- How frequent is this form of addiction and what is the ratio of recognized and unrecognized prescription drug addiction?
- Are those with easier access to prescription drugs (physicians, medical nurses, pharmacists) more susceptible to such addictions?
- How do we recognize prescription drug addiction?
- What is collegiality: honest help or the "this is not my problem" attitude?
- Those struggling with prescription drug addiction frequently obtain prescriptions from multiple physicians. Would it be possible to discover such cases of addiction earlier if the health insurance company informed the personal physician and requested an explanation whenever a patient reached a drug-specific upper limit of consumption in a year?

22.44 Literature as a Medication

A significant proportion of a physician's work is dedicated to psychosomatic problems. Frequently, unfortunate events such as conflicts or failure at work, unemployment, divorce, or loss of a family member lead to anxiety, melancholy, insomnia, and other typical psychosomatic symptoms. The standard response of physicians is to prescribe tranquilizers, sleeping pills, and antidepressants. However, these medications have side effects and can lead to addiction. In addition, they come with significant costs over long-term use.

For such patients, physicians in New Zealand can prescribe a book instead of medications, with the expenses covered by the national health insurance fund. The physician has a list of "uplifting" books and chooses the one that fits the patient best. Similar plans are being considered in Denmark and Great Britain.

Task
Read the book "A Charming Mass Suicide" by Aarto Paasalinna. Would this book be suitable for patients with depression? Present the concept of "literature as a medication." Would it be appropriate to implement something similar in your country?

22.45 Sinclair Lewis: "Arrowsmith"

The novel "Arrowsmith" by Sinclair Lewis is an interesting story about a physician faced with a plague epidemic on a Caribbean island. To prove the benefits of vaccination, he is planning to run a clinical trial with randomized choice ("a randomized trial") where one half of the islands' inhabitants will receive the vaccination.

Task
Read the novel. Present the story and focus on the physician's dilemmas in planning the study—probably the first randomized trial in the history of medicine.

22.46 Female Genital Mutilation

The Nasri family moved to London 10 years ago. The Nasris are originally from southern Egypt at the border with Sudan. They came with a boy who is now 12 years old. Soon after their arrival to Great Britain, their daughter Fadila was born. The second daughter Amira was born 2 years later.

Three years ago, they visited their relatives in Egypt. Upon their return to London, the mother and the elder daughter visited the family physician because Fadila had urine incontinence. The physician learned that her genitals were mutilated ("female circumcision") during the visit to Egypt. The mother defended the procedure for it should supposedly ensure a happy family life to her daughter. This year, the physician was informed by the social services that the family was planning another visit to Egypt. It seems very likely that they will use this occasion to mutilate the genitalia of the younger daughter.

The physician called the mother for a conversation. She assured them that they would not circumcise the younger daughter. The physician and the social worker do not believe her.

Task

Present facts about female genital mutilation (FGM). Search the web for articles with that keyword. Show the scope of the problem and the health consequences of FGM in girls. Questions:

- How should the physician decide when he or she is caught between the respect for cultural identity and diversity and the protection of children's rights?
- Would it be appropriate if the physician violated the principle of confidentiality and reported the suspicion of the intended criminal act to the police?
- How broad is the scope of our ethical duty to act against such practices—does it stop with immigrants living in our countries or is it our duty engage in the prevention of such practices at their roots, that is, in countries where circumcision affects the majority of the female population?
- What is the position of the World Health Organization?

22.47   Accusation of Medical Malpractice

Dr. Sreten Nakićenović is a physician working at the Vrhnika Healthcare Centre. He received a visit by parents who brought in their 6-month-old daughter, Katarina. The girl had recently undergone surgery due to a congenital defect—doubled intestines. After surgery, she became poorly responsive and started vomiting. The physician did not refer the girl to the emergency department; instead, he ordered the parents to monitor their daughter's condition and to come back if it worsened. After a few hours, the family returned with the girl in very poor condition. She was immediately transferred to the emergency department. Shortly afterwards, the girl passed away due to bowel infarction.

The court convicted the physician of professional malpractice, but he was acquitted after appealing the verdict at the court of appeal. The court of appeal did not question the occurrence of professional malpractice; however, in its judgment, it referred to the expert opinion that stated that it could not be ascertained whether the girl would have survived even if Dr. Nakićenović had referred her to the emergency department immediately after the family's first visit.

Task

Present details of the case. Questions:

- Present the arguments in favor of both the guilty verdict and acquittal.
- Following this logic, can a physician ever be found guilty at all? Any disease can end tragically despite the best treatment. This means that regardless of how inappropriate the chosen treatment was, the disease always plays its part and medical malpractice is not the only factor that can be associated with a negative outcome.

22.48 Transport of a Dying Chronic Patient to the Emergency Department

Georgij is a 67-year-old retired construction technician originally from Macedonia. He has been living in Slovenia for 44 years with his family (wife, three children, and five grandchildren). A year ago, he was diagnosed with lung cancer. He received treatment with radiotherapy and chemotherapy. His condition was stable for half a year, after that the breathing difficulties recurred. The attempt of treatment with targeted drugs was not successful. Oncologists informed the family that there was no hope of improvement and that supportive care was the only viable option. At home, Georgij had an oxygen concentrator and was regularly visited by his family physician and home care nurse. He was very weak and could not get up from his bed; he developed clear signs of cachexia. At a height of 175 cm, he only weighed 51 kg.

On October 30, on a Saturday afternoon, his breathing difficulties became worse: he was rasping, his lungs made wheezing sounds, and he was in severe distress. Because their family physician was not available, in a panic the relatives called the emergency medical service. The physician who was on duty decided to transport the patient to the emergency department of the university hospital. Upon their arrival at the department, the patient was already unconscious and was breathing irregularly. He received ampoule of morphine after which he calmed down and 2 h later passed away.

Task

Visit the emergency department and ask the head physician about patients with chronic terminal diseases who are brought to the department in their final days or hours of life. Ask for his or her opinion about the possible measures to prevent this inhumane treatment. Emphasize the importance of communication between the specialist, family physician, and relatives in the light of reducing the number of such cases to the greatest extent possible.

22.49 Dr. Catherine Hamlin

In 2014, Ethiopia nominated the gynecologist Dr. Catherine Hamlin for the Nobel Peace Prize.

Task

Present Dr. Catherine Hamlin and her work in relieving the burden of postpartum injuries in the poverty of Ethiopia. You can find many references on the internet. One of the more interesting ones is:

http://www.nytimes.com/2014/02/06/opinion/kristof-at-90-this-doctor-is-still-calling.html?_r=0

Questions:

- Who is Dr. Catherine Hamlin and what earned her the nomination?
- Present her work. Emphasize the considerable differences between medicine in developed countries and medical problems in developing countries. What is her life path proving to us and what can we learn from it?
- Is top medicine only what they do in famous American and European hospitals or is top medicine also represented in the work of Dr. Hamlin?

22.50 An Aggressive Patient

Many physicians who work directly with patients have experience with aggressive patients or their relatives. Most commonly, these persons are deeply convinced that their rights can only be claimed with threats, intimidation, or openly violent behavior.

Task
Every group member should visit one or two physicians—his or her personal physician or a specialist in any branch of medicine. Ask them to describe cases of aggressive behavior in their patients or the patients' relatives. Present the individual cases in an anonymous form. Questions:

- How should the physician respond in such cases?
- How may physicians protect themselves, their co-workers or their families?
- Is a physician required to treat an aggressive patient?

22.51 Legalization of Marihuana

Some countries have already legalized the use of cannabis for medical purposes. In recent months, the initiatives for legalizing the use of marihuana for "recreational" purposes have gained growing support.

Task
Present marihuana in relation to a spectrum of legal and illegal drugs. Present facts about the medical use of marijuana. Questions:

- What would be the consequences of legalizing marihuana?
- What are the advantages and what the risks?

22.52 Lay People's Attitudes Towards Euthanasia

In public and media discussions of euthanasia, three questions that should be addressed separately are frequently conflated. These three questions pertain to: the right of every individual to a death with dignity, without pain and suffering;

withdrawing of futile intensive treatment in the final period of life, which is misleadingly termed "passive euthanasia"; and the attitude towards active euthanasia, either as a medical measure or as assistance in suicide.

Task

In the seminar, we will try to determine to what extent people without medical education distinguish these three questions. I propose that you carry out a short survey (not more than one page long) among the students of the Faculty of Arts; however, you can also include a different group of respondents. In the first question, you will, therefore, ask about the right to death with dignity and without suffering. As for the question of futile intensive treatment, it is perhaps best to provide a concrete example:

> Marinka is 53 years old and suffers from a very advanced stage cancer. It has spread to all the organs in the pelvic cavity and has caused blockage of both ureters. The physicians' opinion is that the disease progression cannot be stopped and that supportive care is the only appropriate measure. Marinka is receiving appropriate medication and feels no pain; her blood tests indicate loss of kidney function. In such a situation, may we say that a decision not to send her to dialysis is passive euthanasia?

The third question: Marinka told her physician she wished to die. Should the physician perform euthanasia?

Ask those among the respondents who answered "yes" to the last question if they were willing to participate in performing euthanasia themselves or whether they believe this task should be performed by a physician.

22.53 Discrimination

The term "discrimination" has a very negative connotation; however, it means nothing more than "distinguishing."

If we had unlimited means (financial, personnel, space and equipment, availability of organs for transplantation), decision-making would be simple: every patient would receive the treatment that benefits him or her the most. However, because our possibilities never are and never will be unlimited, we must decide to whom to give priority:

- To a younger patient. All other disease outcome predictors being equal, is age an appropriate criterion for choosing an expensive treatment?
- To a female patient with underage children (as opposed to a woman without children).
- To a famous sportsperson (as opposed to an anonymous patient).
- To a physician.

Task

Present priority criteria for organ transplantation. Since these criteria were widely discussed and published, they may be regarded as legally binding. Questions:

- Is it ethically acceptable to use additional criteria when deciding about priority access to a treatment?
- Is any discrimination by default a negative one?
- When is discrimination ethically admissible and when not?

22.54 Is Pedophilia a Disease?

The term "pedophilia" literally translates as "love for children." Is it appropriate that we use this at-first-glance good-natured coinage when talking about the sexual abuse of children?

Task
Read the chapter on pedophilia from the book "The Medicalization of Everyday Life" by Thomas Szasz. Present the societal attitudes towards "abnormal" forms of sexuality. Where to place pedophilia? How are we to understand the practice of sexual satisfaction with a child: as disease or as crime?

22.55 The Nuremberg Trial Against Nazi Physicians

The most famous trials against physicians took place between December 1946 and August 1947 in Nuremberg.

Task
Prepare a presentation about the trial. Questions:

- What crimes were the Nazi physicians accused of?
- Could the other, winning side be convicted of similar crimes?
- Present the Nuremberg Code, which is the first modern document that laid out the rules of medical research.

22.56 Humor in Communication with Patients

Communication between the physician and patient is frequently characterized by tension, unease, and fear. This is particularly burdensome when dealing with serious diseases. In such situations, the words of the physician will not be understood properly. Fear is a great obstacle to channeling empathy and reduces the patient's motivation for treatment.

To improve communication and overcome tension, some physicians resort to serious or humorous comparisons, sometimes even dark humor. For example:

The physician is explaining to the patient why he would recommend radiotherapy after surgery. "Imagine that you removed weeds in your garden with shears: the weeds are seemingly gone, however, in a couple of weeks' time the garden will be again covered in weeds. Although we see no weeds right now, we must nevertheless remove their roots as well."

The patient asks whether he is going to survive. The physician turns to the nurse saying: "Nataša, please provide the gentleman with a five-year guarantee and tell him that he should complain in person in case he dies earlier."

Task

Present the basic guidelines for communication with patients. Questions:

- Emphasize the importance of communication as a gradual building of a relationship.
- Where are the boundaries between acceptable humor, which benefits communication, and jokes that can be perceived as insulting by the patient?
- I am not sure whether the following is true, but my impression is that surgeons (perhaps due to their stressful profession) are the biggest jokers. Meet with some physicians, also surgeons, and ask them to tell you their stories.

22.57   Obamacare—American Healthcare Reform: Successes and Difficulties

The healthcare system in the USA is considerably different from most other healthcare systems in the developed world, Europe, also Canada. Even though the USA is one of the richest countries and that they spend significantly more money on healthcare than most other countries, a considerable part of the population has no health insurance.

Task

Present the current state of healthcare services in the USA and their plans to ensure at least basic healthcare to all their citizens. Questions:

- What are the reasons for the very high costs of healthcare in the USA?
- What is supposed to change?
- What are the reasons and hurdles that will make the goals of the healthcare reform hard to achieve?

22.58   Child Abuse

Children are the most vulnerable members of our society. Child abuse appears in different forms—physical maltreatment, exposure to emotional stress, abandonment, hunger, unhealthy diet.

Task

In your presentation, focus on the ethical dilemmas of the physician who must protect the child but at the same time remain aware of accusations from the parents' side. Questions:

- Present the forms of child abuse.
- When do we suspect that an injury is not a result of an accident but is due to violence?
- The principle of confidentiality versus the interest of the child.
- The importance of precise documentation.
- Suspicion of maltreatment in emergency departments where the physician often does not have an insight into the overall family profile.
- The family as a whole and the necessity of a multidisciplinary treatment.

22.59 Ethical Questions in Self-Inflicted Diseases

Branko is 62 years old and has been receiving treatment for increased blood pressure, diabetes, and increased cholesterol levels. He is 172 cm tall and weighs 115 kg. For a long time, his personal physician has been trying to convince him to quit smoking but to no avail. A month ago, Branko had a myocardial infarction. The physician presented him with a choice: either Branko quit smoking, or he should choose a new personal physician.

Task

Present unhealthy lifestyle as an important factor in the etiology of many diseases. Alcoholism, smoking, drugs, inappropriate dietary habits, severe physical inactivity, unprotected sex with unknown partners, risky behavior in traffic and sports— these are all factors contributing to an increased risk of disease and physical injury.
  Questions:

- Has the physician the right to decline non-urgent treatment of a patient who does not follow the instructions?
- Has the insurance company the right to charge additional fees on top of the insurance premium for individuals who willfully and against all advice harm themselves with unhealthy lifestyles?
- Visit your personal physician and ask him or her what course of action she or he takes when faced with patients who do not follow instructions.
- How does this question relate to the respect for the ethical principles of autonomy (of the patient whose freedom we are trying to limit), beneficence, and justice (with respect to persons who lead a healthy lifestyle but pay the same amount of insurance premiums)?

22.60 Homeopathy

Treatment with homeopathy has a long history and goes back to the days when the drugs of "official" medicine were very frequently inefficacious and often even harmful. After decades of outright rejection, attitudes of physicians towards homeopathy are changing. Some still reject homeopathy as pure charlatanism. However, one can also hear opinions that with certain diseases a placebo is well worth a try. In all this, the physician is the only person who can safely use homeopathic drugs: he or she knows how to establish accurate diagnosis, knows when it is appropriate to try a placebo, and also knows when one needs to switch back to the treatment recommended by "official medicine."

Task
Present the basic postulates of homeopathy, its history and its present position. Questions:

- Can you find a methodologically correct, unbiased, double-blind, randomized trial which would prove the efficacy of homeopathic drugs against placebo (pure water)?
- Are there medical cases you would entrust to homeopathic drugs?
- If so, what are these areas and what are the safeguards we must follow not to harm the patient?
- How is the field of homeopathy regulated by the legislation in individual member states of the EU?

22.61 Treatment of the Demented Patient

Milena is 84 years old and is in nursery care. Due to senile dementia and her post-stroke condition she is completely immobile, she is fed through a nasogastric probe. She does not talk and cannot recognize her relatives. Treatment is led by the nursing home physician. Last year, Milena recovered from pneumonia. During a recent febrile episode, the physician established the recurrence of pneumonia and prescribed antibiotics.

Her son (himself a physician but not the personal physician of his mother) questions the appropriateness of treatment with antibiotics. He never discussed the meaning of life with his mother nor did they talk about her wishes in situations of permanent loss of communication and complete dependence on assistant care. However, if he himself had ended up in such a state, he is certain that he would not want to have life prolonged, regardless of what treatment is chosen for the purpose.

Task

Present ethical issues concerning futile intensive treatment. A similar case is presented in the *New England Journal of Medicine*: http://www.nejm.org/doi/full/10.1056/NEJMclde1411152?query=TOC

Question:

- Should the son propose to his colleague, the nursing home physician, to withdraw the antibiotic treatment of pneumonia?

22.62   The Physician in Commercials

Damjan is a handsome physician specializing in orthodontics. As a student, he took part in a film as a background actor. A television producer is now offering him a role in a toothpaste commercial.

Uršula is a family physician, responsible for a region with many mountain hamlets. When she is buying a new car, the car salesman offers her a special discount if she was willing to step in front of the camera for promotional purposes and say: "I could not imagine doing my work without a reliable *HipHop* SUV."

Tamara works as a private physician in a dermatology office. She gave a long interview for *Forever Young* magazine. In the interview, she emphasizes that *Ultraskin* body cream is the best cream for healthy and youthful skin.

Task

Present the position of the Code of Medical Ethics regarding the participation of physicians in commercial advertising. Questions:

- Is it admissible for physicians, as known representatives of their profession, to take part in commercials?
- Is the "selling" of one's profession really unacceptable, humiliating or is it just the envy of those colleagues who do not do it themselves? Are physicians any different from athletes who successfully capitalize on their successes?

22.63   The Ebola Epidemic: Ethical Questions

The Ebola epidemic took tens of thousands of lives in Africa and only sporadic deaths in a few European countries and the USA.

Task

Present the basic facts on the Ebola epidemic. Focus on how the West (Europe, USA) responded to the epidemic: governmental and non-governmental organizations, the media, pharmaceutical industry. Dr. Iza Ciglenečki is a Slovenian physician working for the non-governmental organization Doctors Without Borders. Ask her to describe her experience with the Ebola epidemic. Question:

- What was the Western response focused on—the epidemic alone, the victims of the disease and the suffering of the local peoples, or the fear of disease spreading beyond Africa?

22.64   Gene Testing in Underage Daughters

At the age of 36, Marija was diagnosed with breast cancer. Both her mother and her sister have been treated for breast cancer as well (both are healthy now). Given Marija's young age and the two cases of breast cancer in close family members, oncologists began suspecting a genetic predisposition for the disease. Marija agreed to gene testing: the BRCA-1 gene mutation was confirmed. Marija has two daughters, 9 and 11 years old. Marija is informed that for each of her daughters there is a 50% chance that they inherited the mutant gene and that approximately 80% of women with the BRCA-1 gene mutation develop breast cancer between 25 and 70 years of age. Marija wants immediate gene testing for her daughters.

Task
Present legal regulation of gene testing and practice, especially for underage persons. Questions:

- Should the oncologist perform testing in underage daughters?
- Present ethical analysis of gene testing for underage persons, with the related ethical benefits and burdens.

22.65   Revocation of Driver's License

A physician works in a small town (healthcare center, three family physicians) where everyone knows each other. One of her patients is Matija, a 60-year-old chronic alcoholic, who refuses to quit alcohol and rejects treatment but, nevertheless, still drives a car. He has already had some minor road accidents. The physician is worried for the safety of her children and of the other town residents and considers sending a report on his incapacity to drive to the police. However, due to the close personal relations in the community, it is likely that Matija will learn about her report and might threaten her.

Task
Present the legislation on issuing/validity/withdrawal of the permit to drive. Questions:

- Talk to at least three family physicians (of your own choice) and ask them how they would act in such a situation.
- Should the physician report about Matija's inability to drive a car to the traffic police and suggest the revocation of his driver's license?

- Ethical analysis for the decision to send a report to the authorities: prepare a list of persons involved, with their related benefits and burdens according to the four ethical principles.

22.66 Airplane Seats for Overweight Persons

On long flights, an airline offers wider, but more expensive seats for travelers weighing more than 120 kg. According to the new proposal, travelers should indicate whether their body weight is above the limit when buying a ticket. The move was spurred by the complaints of passengers who had to sit between two overweight passengers. An association of diabetes patients brought a complaint to the ethical committee claiming the offer of the airline is discriminatory against those weighing over 120 kg because they have to pay more for their ticket.

Task
Go to the web to find examples of problems with respect to heavily overweight travelers in air transport. Questions:

- Ethical analysis: the new proposal for wider seats, with benefits and harms for passengers of average weight and for overweight passengers.
- Discuss the issue of protecting privacy: data on overweight passengers might be misused.
- In conclusion, should the ethical committee support the complaint of the diabetes patients' association?

22.67 Individual Consent for Review of Old Biopsies

The Department of Pathology of the Institute of Oncology in Ljubljana keeps archives of tumor biopsies acquired over the past 60 years. Researchers intend to apply a new technique to study biopsies of patients with soft-tissue sarcomas who were treated at the institute in previous decades. The objective is to find correlations between the pathological profile of the tumor and the clinical course of the disease. The great majority of biopsies were taken at a time when patients did not specifically consent to the use of their biological material for research unrelated to their treatment. There are no commercial sponsors involved in the project.

Should the Ethical Committee require the consent of patients, or (in case they are no longer alive) of their relatives before a review of the biopsies is performed?

Task
Review the legal regulation for patient's consent in Slovenia and the EU. Present difficulties for obtaining consent from patients treated a long time ago, and especially from their relatives. Furthermore, if studying only biopsies with consent, the analysis may be biased since consent is especially difficult to obtain from patients who died.

Question:

- Should the relevant ethical committee require that the patients (or their relatives if the patient passed away) give individual consent for the re-analysis of biopsies?

22.68 Love in a Nursing Home

Martin is 81 years old and has senile dementia. He left his home several times and got lost. At 73, his wife Elizabeta cannot take care of him. Martin was admitted to a nursing home where he fell in love with another home resident, a widow who struggles with walking but is otherwise in good mental condition. They hold hands and spend entire days in each other's company; however, Martin does not recognize his wife anymore. Elizabeta is deeply shocked. Martin's and Elizabeta's children request the administration of the nursing home to move Martin to a different building where he would have no contact with his new partner. If the administration is not willing to do so, the children will take their father to a different nursing home.

Task
Visit a nursing home and ask about emotional relationships between clients. Questions:

- Ethical analysis: who benefits and who is hurt if the demand for moving Martin to another location is granted?
- May we expect a similar story to be repeated at another location?

22.69 Professional Sports in Children

A naive television sports spectator imagines professional sports to be characterized by an abundance of health. The truth is much crueller: many professional athletes will end up disabled in their mature age. This seminar focuses on the consequences of professional sports in children. The question is crucial because it holds in almost all competitive sports that only the competitors who started with intensive training early in their childhood can achieve top results.

Task
Present the consequences of excessive physical activity for the physical and mental development of children. Questions:

- Where is the line between healthy physical activity and competitive sports in children?
- How should we distinguish between healthy ambitions in the young and the abuse of the child when he or she is pushed into the sport by their parents or coaches with their unfulfilled ambitions or even financial interests?

22.70 Medical Strike

Without a doubt, we physicians have the right to demand adequate conditions for our work: the ability to pursue diagnostics and treatment according to the best interest of our patients, safe working environment, and fair payment. When some of these conditions are not met, physicians may consider a strike.

Task
Present legal and ethical limitations to medical strike. Questions:

- What are the differences between the strike of industrial workers and that of physicians?
- When is a medical strike justified?
- What ethical dilemmas open up when physicians announce a strike or when they start it?
- How should a medical strike be carried out?

22.71 Artificial Womb

As a member of an ethical committee, you are reviewing a project proposal for the development of an artificial womb. Researchers plan to implant a fertilized egg into an artificial womb which would simulate the conditions within a real one: a bed for embryo's implantation, oxygenation, supply of nutrients, temperature, movement, circadian rhythm. The authors of the proposal ensure this would allow for genuine gender equality because even single men and homosexual couples would be able to have children. Furthermore, an artificial womb would also mean that women in heterosexual couples would be relieved of the trouble and risks of pregnancy and childbirth.

Task
Present some historic attempts to construct an artificial womb. Questions:

- Where can you see the benefits and burdens of such a device?
- Will you approve or reject the proposal?

22.72 Genetic Testing for Prediction of a Disease

At 39, Jennifer is perfectly healthy. Because her mother developed breast cancer at an early age, Jennifer decided to undergo testing for two genes (BRCA 1 and BRCA 2), which can be predictors for a high risk of developing breast cancer. A consultant from the genetic information service suggested that she undergo testing for 20 other cancers and Jennifer agreed.

The results were striking. While both breast cancer genes were negative, the tests revealed a mutation of a gene that predicts a high chance of stomach cancer. In persons who test positive and who have a blood relative diagnosed with stomach cancer, the risk of stomach cancer is so high that it is recommended to remove the stomach entirely as a preventive measure. No data are available, however, for the risk of stomach cancer if only the test is positive without records of disease in the family.

Task

Read the New York Times article:

http://www.nytimes.com/2014/09/23/health/finding-risks-not-answers-in-gene-tests.html?_r=0.

Questions:

- Why is professional counselling an essential part of genetic testing?
- Ethical benefits and burdens of genetic testing.

22.73 Cancerphobia

Martina is 53 years old, divorced, and has two children. Last year, her sister was diagnosed with breast cancer. Years ago, her stepbrother died of lung cancer.

Martina read several books on cancer. She is convinced that she is also getting cancer and would see any, even the tiniest health issue as the first sign of cancer. Every year, she undergoes breast screening with mammography and blood testing for tumor markers. Last year, she had mild digestive problems. At her insistence, her physician referred her to examination with a colonoscopy, which did not reveal any sign of disease. Because her physician did not believe that she had cancer, Martina decided to change her personal physician. After she read on the internet that a PET/CT scan can show where cancer is developing in the body, she demanded that her new physician refer her to such an examination. The physician refused to issue a referral for a PET/CT scan for which there was no medical indication.

Task

Visit your personal physician and ask him about patients with cancerphobia. Questions:

- How should a physician act when a patient requires diagnostics for which there is no indication?
- The statistics tell us that every third woman, and every second man will develop one of the 100 different forms of cancer. With this in mind, how can a physician protect him or herself from the accusations of negligent examination if one day the person with cancerphobia indeed develops cancer?

22.74   Communication in the Waiting Room

When discussing communication, we are almost exclusively focused on the communication between the physician and other healthcare workers on the one hand and the patient and his or her family on the other. Communication between patients receives considerably less attention. Especially noticeable is the communication between chronic patients who are frequently returning for treatment, share their experience, recommendations, and make friendships.

Task
Review the literature on communication. For practical experience, visit the office at the university clinic where they treat chronic patients. Engage in a conversation with the patients and prepare a presentation on what they said.

22.75   Traditional Medicine

Traditional medicine with its customs, recipes, and advice is an integral part of the ethnological heritage of many nations. In Slovenia, Carinthian teacher, writer, and politician Vinko Möderndorfer was collecting such resources for several decades. In 1964, the Slovenian Academy of Sciences and Arts published his book on traditional medicine with advice on the use of medicinal herbs and with folk superstitions, spells, and other measures that our ancestors used to chase away diseases. Students from other countries should find published work on folk medicine in their nation.

Task
Present a selection of traditional folk recipes and advice and provide comments from the perspective of contemporary medicine.

22.76   Late Termination of Pregnancy

When learning about the fascinating reports and the publication of Slovene scientists on their evidence of the Zika virus in the brain of an unborn child, some of us became aware of the fact that a pregnancy was terminated in the 32nd week.

Task
Read the article at the following link: http://www.nejm.org/doi/full/10.1056/NEJMoa1600651?query=featured_home#t=articleTop.
   Question:

- What are the ethical and legal provisions on late termination of pregnancy, that is, the termination of pregnancy after the 10th week?

22.77   Postponement of Prison Sentence Service for Health Reasons

One of the tricks used by the convicted is also to postpone their prison sentence term for health reasons. After the sentence is final, certain influential convicts manage to avoid serving their prison term for a longer period or even forever.

Task
You will prepare this seminar together with the students from the Faculty of Law. Law students should highlight the legal provisions that allow postponement of prison sentence service for health reasons. Medical students will emphasize the role of the physician: under what conditions is it legally and ethically acceptable for the expert physician to sign the opinion that the convict is not capable of serving the sentence for health reasons.

I offer a provocative standpoint: every convict's health condition deteriorates while serving a prison sentence—if we understand "health" (following the definition of the World Health Organization) as a state of perfect physical, mental, and social well-being. A prison term is simply not the same as a vacation in Hawaii.

Questions:

- In every case, a physician may conclude that the prison sentence will deteriorate the convict's health condition. Is such an opinion justified?
- Should the judge follow such an opinion of the physician?

22.78   Animals in Biomedical Research

It is not possible to imagine medical research without experiments on animals. Evidently, this also opens certain ethical dilemmas. My personal experience: after the Chernobyl disaster, I was visiting Ukraine with a group of international experts, where they were planning to build a center for research on the effects of ionizing radiation. Among other things, they wanted to build a laboratory for investigating the effects of ionizing radiation on the behavior of chimpanzees.

Task
Read and present the article: Bateson, P. (2011) Ethical debates about animal suffering and the use of animals in research. *Journal of Consciousness Studies*, **18**, 186–208.

Evaluate the aforementioned project from an ethical standpoint: definition of moral status for all concerned, as well as ethical benefits and burdens.

22.79   Death of Grandparents

In the case of a severely diseased elderly patient, physicians typically talk to his or her adult children and frequently ignore the grandchildren. However, it is the grandchildren who can be most traumatized by the death of their grandfather or grandmother. Their parents might have been hiding the disease of the grandparents, and the death comes even as a greater shock.

Task
Visit the Unit for Palliative Care of the University Medical Centre and ask the physicians about their experience regarding children at the bedside of an elderly patient.
   Questions:

- Any stereotypes on "typical" human relations are problematic. Nevertheless, try to present characteristics of relations between an adult woman and her mother, as compared to a granddaughter and her grandmother.
- How should we talk to children about disease and death?
- Has anyone among you experienced something similar?

22.80 Guerilla Surgeon

In his book "Guerilla Surgeon," Dr. Lindsay Rogers, a physician from New Zealand, describes how the Second World War brought him to the Partisans in Slovenia.

Task
Read the book and give a short presentation. Present specifically the case of a young Partisan who was blinded after a grenade explosion but recovered his sight after Dr. Rogers removed a metal fragment from the brain. The daughter of this Partisan is a physician, now retired, and she will be happy to retell the story of her father. You can also present other testimonies about Partisan healthcare (Franja Partisan Hospital). Can we imagine medicine and the work of physicians in such extreme conditions today?

22.81 Loneliness

In Sweden, every other adult person lives alone, and every fourth person dies alone. Loneliness is becoming a problem in many countries. Especially the elderly and the widowed frequently remain alone.

Task
Talk to at least three family physicians and ask them to give you concrete examples of loneliness, especially among the elderly. Ask them how they go about such issues and how they solve them. Questions:

- What are the consequences of loneliness from a medical perspective?
- How can we mitigate this problem?

22.82 Mental Health of Political Leaders

After Donald Trump's election, it became clear that his (to put it mildly) unusual standpoints are not only part of his election campaign rhetoric. Some are very openly suggesting that he is a dangerously unpredictable personality.

Task
Present the general rules for respecting a patient's privacy. Questions:

- In which situations may a physician break his commitment to confidentiality and reveal information on his patient's health to the public?
- Is it appropriate that a psychiatrist enter public discussions on the mental health of a political leader, while not being involved in the medical care for the persona in question?
- Can you see a difference in the meaning of words such as "paranoiac" or "schizophrenic" when used by laypersons (e.g., members of parliament during a hot political debate), as compared to their use by a psychiatrist (in a public comment on political situation)?

22.83 Eugenics

The new possibilities of medically assisted reproduction pave the way towards embryo selection. These procedures may be used not only for selecting and eliminating serious genetic diseases but also for selecting other characteristics of the child: gender, appearance and physique, mental characteristics, etc.

Task
Give a short presentation of the film Gattaca: https://123movies.ru/film/gattaca-13712/watching.html.
Questions:

- Will we really be having "custom-made" children in the near future?
- What is the line between ethically acceptable and ethically unacceptable?

22.84 Fatherhood

Dejan has been receiving treatment for diabetes since his childhood. At the age of 27, he developed end-stage kidney failure and is listed for kidney transplantation. In the process of finding a suitable donor, a histocompatibility analysis was performed among his parents and close relatives. The physicians established that Dejan's father is not his biological father, while his father's brother is a perfect donor.

Task
Present one of the ethical problems of genetic testing: revealing unsolicited genetic information. Question:

- Should the family be informed about the outcomes of genetic analysis regarding fatherhood?

22.85 Mark Langervijk

In the Netherlands, euthanasia was performed for Mark Langervijk, a 41-year-old chronic alcoholic. The entire story is available here:
   http://www.dailymail.co.uk/news/article-3980608/Dutch-euthanasia-law-used-kill-alcoholic-41-decided-death-way-escape-problems.html.
   Another story: euthanasia was performed for a woman in her 20s who suffered from severe psychological issues due to sexual abuse in childhood:
   http://www.cbsnews.com/news/netherlands-sex-abuse-victim-euthanasia-incurable-ptsd-assisted-suicide/.

Task
Requirements for euthanasia in countries where the procedure is legal. Present the two cases. Question:

- Leaving other pro/contra reasoning aside, can we agree with euthanasia for persons with limited autonomy?

22.86 Molière and the Characters of Physicians

In his works, the famous French playwright Molière portrayed physicians with a very satirical tone. Some of these portrayals remain valid.

Task
Show some of the Molière's portrayals of physicians. *The Doctor in Spite of Himself*, *Don Juan*, and *The Hypochondriac* are just three from among his numerous works that give a critical and satirical treatment of French physicians in the seventeenth century.
   Show some of the typical passages and provide comments.

22.87 Alternative Diagnostics

When talking about alternative (non-conventional) medicine, we most often think of alternative modes of treatment, but rarely of alternative diagnostics. Energy systems in the body, bioresonant diagnostics, karmic diagnostics, visual diagnostics, pulse diagnostics, iridology—this is just a short list of several alternative diagnostic procedures that supposedly can be used for establishing the causes of diseases.

For every diagnostic method, we need to know what is the probability that the method detects (sensitivity) a particular disease (e.g., cancer) and what is the probability that at a later stage this suspicion is indeed confirmed (specificity). How do these notions apply to alternative diagnostics?

Task

Show what is being offered as a substitution for standard medical diagnostics. Present the concepts of sensitivity and specificity and compare standard versus alternative diagnostic methods. Questions:

- Describe positive (if any) and negative aspects of alternative diagnostics.
- What should a family physician do when his or her patient presents a "test result" from an alternative diagnostic method and now demands additional conventional diagnostics for which there is no indication?

22.88 Trade with Human Organs for Transplantation

Due to the lack of organs for transplantation, the international—most often illegal— trade with organs is flourishing. Only a few countries allow the financial compensation of donors (Singapore, Australia, Iran). Organs are obtained from people who due to their personal hardship sell parts of their bodies, such as a kidney, or from the deceased. In some cases, victims are even killed explicitly for their organs. The procurement of organs from a living or a dead donor is obviously not possible without the involvement of physicians.

Task

Present the extent of the organ trade worldwide. Questions:

- Compare the profits of all involved in organ trade to the payment to organ donors and present additional burdens to them (morbidity from the procedure, impaired physical condition).
- How can we ethically assess the activities of physicians responsible for the explantation of organs?
- Should patients be informed about the origin of an organ they are about to receive?
- In many countries (including Slovenia), blood is donated on an altruistic basis and therefore without financial compensation. What might be the benefits and risks if blood donors would receive payment for their act?

22.89   Dr. Thomas Percival: Medical Ethics

In 1803, an English physician Thomas Percival published a book entitled *Medical Ethics, or, a Code of Institutes and Percepts, adapted to the Professional Conduct of Physicians and Surgeons*. This book established Dr. Percival as the father of contemporary ethical thought and served as the basis for the Code of Medical Ethics of the American Medical Association and several other ethical codes and declarations.

Task
Read the book (its pdf version can be found on faculty websites). In the presentation, focus on Chapter II: Of Professional Conduct in Private, or General Practice, subsections I, II, III, and IV (pages 30–33) and the footnotes in this part of the book (Notes VI, VII, and VIII, pages 154–168). Present the advice to physicians regarding alcohol/sobriety. Present Thomas Percival's advice on the question of communicating the truth to the severely ill patient in situations in which there is a conflict between telling the bitter truth or maintaining a false, unreal hope. How are these questions addressed today?

22.90   Mielke and Mitscherlich: "Doctors of Infamy—The Story of the Nazi Medical Crimes"

The authors are providing detailed documentation of the crimes committed by German physicians during WWII. The book is based on the evidence and hearings of the accused during the Nuremberg trials that took place between October 1946 and August 1947. The incriminating medical experiments were designed so as to provide knowledge that would benefit German soldiers in life-threatening situations, such as being shot down from an aircraft at high altitudes, severe hypothermia, or typhus infection. Certain crimes, however, had absolutely no potential practical use and reflected pure sadism. An example is the collection of human skulls of different nationalities and races for the medical faculty in Strasbourg. As the human "material" in experiments, they used the prisoners of "inferior" races (Jews, Poles, Gypsies) from the concentration camps.

In 1949, the book was printed in 10,000 copies in order to be freely distributed among all German physicians, but it never reached the intended audience.

Task
Present some of the experiments: on survival at high altitudes, on supercooling, on vaccination against typhus, on bone transplantation, the euthanasia plan. Questions:

- What was the role of the Nazi physicians in designing the experiments? May we accept the argument of defense at the trial that all experiments were approved by the highest state administration led by Himmler?
- Present the facts about the mass collaboration of German physicians with the Nazi regime. Since WWII, German medical associations have not explicitly

investigated or have not distanced themselves from the past crimes. For further information, see the paper by Pross Ch. Breaking through the postwar cover-up of Nazi doctors in Germany. J Med Ethics 1991; 17: Suppl 13–16 (http://jme. bmj.com/content/medethics/17/supplement/13.full.pdf).

22.91   Oregon: The Death with Dignity Act

In 1997, the US state of Oregon was one of the first to pass a law giving the terminally ill the right to "death with dignity." It allows physician-assisted suicide, most frequently in the form of obtaining a prescription for a lethal dose of sleeping pills (barbiturates).

Task
Present the experience with the enforcement of the law, the arguments in favor of the law, and the arguments against it. Give a clear overview of abuses when patients should receive a different form of help rather than a lethal dose of sleeping pills.

Printed in the United States
By Bookmasters